Television Microprocessor IC Data Files

Television Microprocessor IC Data Files

John Edwards

Newnes
An imprint of Butterworth-Heinemann
Linacre House, Jordan Hill, Oxford OX2 8DP
225 Wildwood Avenue, Woburn, MA 01801-2041
A division of Reed Educational and Professional Publishing Ltd

A member of the Reed Elsevier plc group

OXFORD BOSTON JOHANNESBURG
MELBOURNE NEW DELHI SINGAPORE

First published 1997
Reprinted 1998

British Library Cataloguing in Publication Data
A catalogue reference for this book is available from the British Library

Library of Congress Cataloguing in Publication Data
A catalogue record for this book is available from the Library of Congress

ISBN 0 7506 3335 2

Printed and bound in Great Britain by
Biddles Ltd, Guildford and King's Lynn

Preface

As domestic home entertainment equipment such as television receivers, video recorders and hi-fi systems become more and more sophisticated but cheaper to purchase, the consumer electronics engineer must be armed with a high degree of technical knowledge and quick access to service information relating to the equipment under repair. It is the latter in which the service engineer is badly served by the industry. A fast economic repair is the only way a repair business can remain viable. The majority of manufacturers technical departments will not supply technical assistance to a non account engineer struggling with a nasty fault symptom. Service manuals are usually too expensive, no longer available or poor quality photocopies and all too often the information they contain do nothing in the way of explaining how the circuits function.

All modern electronic equipment use dedicated integrated circuits. In this book you will find the most common devices encountered in the day to day repairs of television receivers. The signal routes in and out of the devices are also shown, the measurements were taken under actual working conditions. Where it's considered helpful, voltage readings, also taken under actual working conditions are included.

This book is not intended to explain the intricate internal workings of the devices – that would in any case be of little use in fault finding; also some measurements will vary depending on the associated circuitry and test equipment used. It is also true that some devices can be used in various ways, but within these pages are the most common applications you will encounter and only minor variations will be found.

This book was written by a practising engineer of over 25 years' experience as an aid for the engineer of today. This first and future editions of Television IC Data Files will build up to a valuable reference source, taking the frustration out of not knowing what signal should be where during fault finding.

Note: DIL IC diagrams are shown as seen from the top, and SIP diagrams from the side that shows the pin 1 identification. Voltage figures are DC, waveforms are pk-pk. Devices that use the Greek μ designation (such as μPD1514C) are listed under U (as UPD1514C, for example) because both letters are used interchangeably in listings.

BU2458-09

Infra-red remote controller IC

Pins
(1) N/c
(2) Ground
(3) ⟶ VCR/TV switch
(4) ⟷ Keypad functions scanning
(5) ⟷ Keypad functions scanning
(6) ⟷ Keypad functions scanning
(7) ⟷ Keypad functions scanning
(8) ⟷ Keypad functions scanning
(9) ⟷ Keypad functions scanning
(10) ⟷ Keypad functions scanning
(11) ⟷ Keypad functions scanning
(12) ⟷ Keypad functions scanning
(13) ⟷ Keypad functions scanning
(14) ⟷ Keypad functions scanning
(15) ⟷ Keypad functions scanning
(16) ⟷ Keypad functions scanning
(17) ⟷ Keypad functions scanning
(18) ⟷ Keypad functions scanning
(19) ⟶ Infra-red transmitting diode driver
(20) ⟷ System oscillator 393 kHz
(21) ⟷ System oscillator 393 kHz
(22) ⟵ 3 V supply

Notes:

CCU2070

Control microprocessor IC

Pins

(1) ←→ System oscillator 4 MHz
(2)
(3) → 976.6 Hz clock external EEPROM memory
(4) ← Power on reset = low
(5) ← Power switch pulse contact, make = low
(6) Chassis
(7) ←→ Data IM bus
(8) → Ident IM bus
(9) → Clock IM bus
(10) → Channel tune up
(11) → Channel tune down
(12) ← Remote control data
(13) ← Tuner pre-scale oscillator sample
(14) ←→ Keypad function scanning
(15) ←→ Keypad function scanning
(16) ←→ Keypad function scanning
(17) ←→ Keypad function scanning
(18) ←→ Keypad function scanning
(19) ←→ Keypad function scanning
(20) Chassis
(21) ←→ Keypad function scanning
(22) ←→ Keypad function scanning
(23) N/c
(24) N/c
(25) N/c
(26) ← Service mode on = low
(27) ← 5 V supply
(28) ← SCART socket status pin 8
(29)
(30)
(31) → Received transmission system select
(32) → TV/external select
(33) → Tuning band VHFL, VHFH, UHF select
(34) → Tuning band VHFL, VHFH, UHF select
(35) → Tuning band VHFL, VHFH, UHF select
(36) ←→ Keypad function scanning

(Continued opposite)

(37) ←——→ Keypad function scanning
(38) ←——→ Keypad function scanning
(39) ←——→ Keypad function scanning
(40) ←—— 5 V supply

Variation

Pins
(2) Chassis
(31), (32), (36), (37) N/c

....

Notes:

CCU3000

Control microprocessor IC

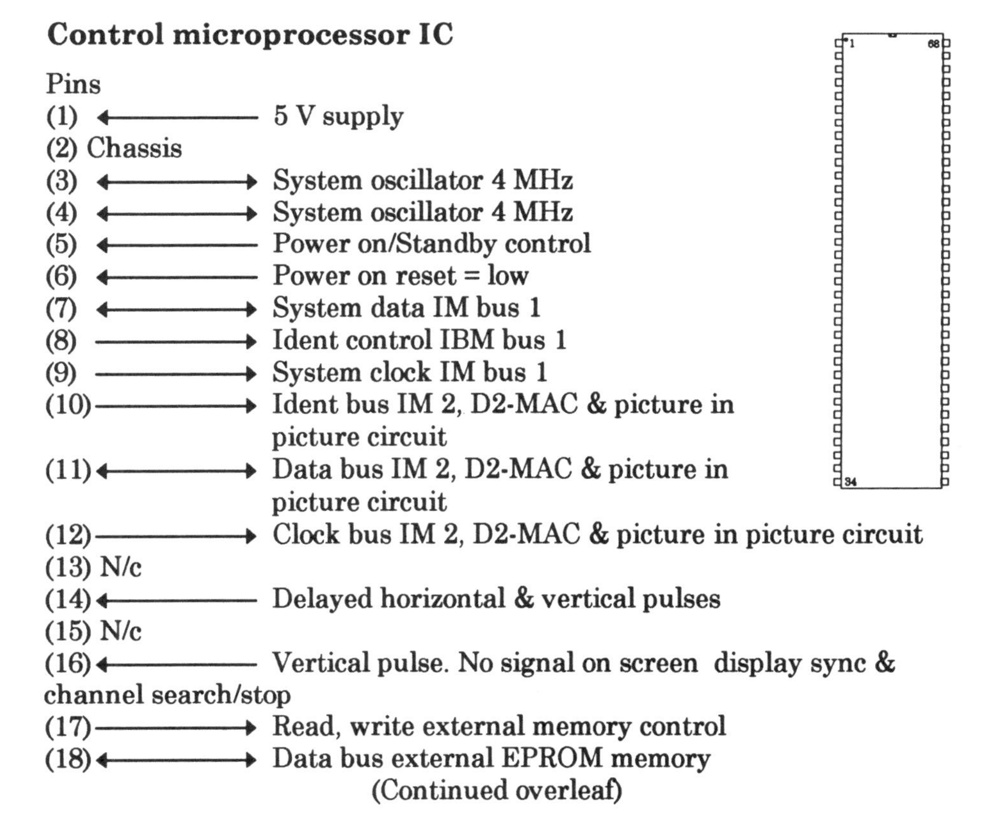

Pins
(1) ←—— 5 V supply
(2) Chassis
(3) ←——→ System oscillator 4 MHz
(4) ←——→ System oscillator 4 MHz
(5) ←—— Power on/Standby control
(6) ←—— Power on reset = low
(7) ←——→ System data IM bus 1
(8) ——→ Ident control IBM bus 1
(9) ——→ System clock IM bus 1
(10) ——→ Ident bus IM 2, D2-MAC & picture in picture circuit
(11) ←——→ Data bus IM 2, D2-MAC & picture in picture circuit
(12) ——→ Clock bus IM 2, D2-MAC & picture in picture circuit
(13) N/c
(14) ←—— Delayed horizontal & vertical pulses
(15) N/c
(16) ←—— Vertical pulse. No signal on screen display sync & channel search/stop
(17) ——→ Read, write external memory control
(18) ←——→ Data bus external EPROM memory

(Continued overleaf)

CCU3000

(19)←——→ Data bus external EPROM memory
(20)←——→ Data bus external EPROM memory
(21)←——→ Data bus external EPROM memory
(22)←——→ Data bus external EPROM memory
(23)←——→ Data bus external EPROM memory
(24)←——→ Data bus external EPROM memory
(25)←——→ Data bus external EPROM memory
(26)——→ Address bus external EPROM memory
(27)——→ Address bus external EPROM memory
(28)——→ Address bus external EPROM memory
(29)——→ Address bus external EPROM memory
(30)——→ Address bus external EPROM memory
(31)——→ Address bus external EPROM memory
(32)——→ Address bus external EPROM memory
(33)——→ Address bus external EPROM memory
(34)——→ Address bus external EPROM memory
(35)——→ Address bus external EPROM memory
(36)——→ Address bus external EPROM memory
(37)——→ Address bus external EPROM memory
(38)——→ Address bus external EPROM memory
(39)——→ Address bus external EPROM memory
(40)——→ Address bus external EPROM memory
(41) N/c
(42)——→ Data to SCART socket pins 10 & 12 also external memory, used in service modes & satellite equipped sets.
(43)←—— Vertical scan output circuit failure sensing & excessive beam current, enables standby mode.
(44)←—— SCART socket 1 status pin 8
(45)←—— SCART socket 2 status pin 8 includes 4:3 or 16:9 picture ratio information
(46)←—— SCART socket 3 status pin 8
(47)←—— SCART socket 4 status pin 8
(48) N/c
(49)←——→ IM data bus picture in picture circuit
(50)←——→ IM data bus D2MAC circuit
(51)——→ External address bus EPROM memory
(52)——→ D2-MAC mode output
(53)←——→ 12C data bus 2
(54)←——→ 12C data bus 1
(55)——→ 12C clock bus
(56)——→ External RGB output decoder fast picture blanking

(Continued opposite)

(57)————→ Teletext circuits reset
(58)————→ Sound circuits reset
(59)←———— Satellite tuner
(60)←———— Remote control data
(61)←———→ Keypad function scanning
(62)————→ Power on/standby control
(63)←———→ Keypad function scanning
(64)←———→ Keypad function scanning
(65)←———→ Keypad function scanning
(66)←———→ Keypad function scanning
(67)————→ Standby LED display
(68)————→ Satellite tuner standby LED display

Notes:

CRT150

Control microprocessor IC

Pin number
(1) ————→ Channel tuning
(2) ————→ Picture contrast control
(3) ————→ Picture brightness control
(4) ————→ Picture colour control
(5) ————→ Sound volume control
(6) ————→ Tuning band selection VHFL, VHFH, UHF
(7) ————→ Tuning band selection VHFL, VHFH, UHF
(8) ————→ Tuning band selection VHFL, VHFH, UHF
(9) ←———— Auto fine tuning
(10)←———→ Clock bus & keypad function scanning
(11)←———→ Data bus & external memory circuit
(12)←———→ External memory circuit & keypad function scanning
(13)←———→ Keypad functions scanning
(14)←———→ Keypad functions scanning
(15)←———→ Keypad functions scanning

(Continued overleaf)

CCU3000

(16)←——→ Keypad functions scanning
(17)←——→ Keypad functions scanning
(18)←——→ Keypad functions scanning
(19)←——→ Keypad functions scanning & external memory enable
(20)——→ Timer mode on = low
(21) Chassis
(22)——→ Power on/off to relay driver
(23)——→ Green on screen display
(24)——→ Red on screen display
(25)——→ On screen display picture blanking
(26)←—— On screen display horizontal sync pulse
(27)←—— On screen display vertical sync pulse
(28)←——→ On screen display oscillator
(29)←——→ On screen display oscillator
(30) Chassis
(31)←——→ System clock oscillator 4 MHz
(32)←——→ System clock oscillator 4 MHz
(33)←—— Power on reset pulse = low
(34) Chassis
(35)←—— Remote control data
(36)←—— Received signal ident pulse
(37)——→ TV/AV selection
(38)——→ TV/AV selection
(39) Chassis
(40) Chassis
(41) Chassis
(42)←—— 5 V supply

Variations

(6), (7) N/c
(39)——→ NICAM circuitry
(40)——→ NICAM circuitry
(41)——→ NICAM circuitry

Notes:

CTV322S - V1.2

Control microprocessor IC

Pins
(1) ——→ Remote control data
(2) ——→ Sound volume control
(3) ——→ Picture brightness control
(4) ——→ Picture colour control
(5) ——→ Picture contrast control
(6) ——→ Picture hue control for NTSC reception
(7) ——→ Tuning band selection
(8) ——→ Tuning band selection
(9) ←—— Auto fine tuning control
(10) N/c
(11) ——→ VTR select
(12) ——→ External AV source select
(13) ←——→ Keypad function scanning
(14) ←——→ Keypad function scanning
(15) ←——→ Keypad function scanning
(16) ←——→ Keypad function scanning
(17) ←——→ Keypad function scanning
(18) ←——→ Keypad function scanning
(19) ←——→ Keypad function scanning
(20) ←——→ Keypad function scanning
(21) Chassis
(22) ——→ Red on screen display
(23) ——→ Green on screen display
(24) ——→ Blue on screen display
(25) ——→ On screen display picture blanking
(26) ←—— On screen display horizontal sync pulse
(27) ←—— On screen display vertical sync pulse
(28) ←——→ On screen display oscillator
(29) ←——→ On screen display oscillator
(30) Chassis
(31) ←——→ System oscillator 10 MHz
(32) ←——→ System oscillator 10 MHz
(33) ←—— Power on reset
(34) ←—— Received signal ident pulse
(35) ←—— Remote control data
(36) Chassis
(37)

(Continued overleaf)

CX522-099

(38) Chassis
(39) ⟶ System clock bus
(40) ⟷ System data bus
(41) ⟶ Power on/standby control
(42) ⟵ 5 V supply

Notes:

CX522-099

Control microprocessor IC

Pins
(1) ⟶ Picture mute
(2) ⟶ Power on
(3) ⟶ Picture/sound blanking
(4) ⟶ Standby = low
(5) ⟶ Tuner auto gain control
(6) ⟶ Tuning band selection
(7) ⟶ Tuning band selection
(8) ⟷ Keypad functions scanning & external memory, also games input interface
(9) Chassis
(10) ⟶ On screen display character
(11) ⟶ On screen display picture blanking
(12) ⟶ Auto fine tuning up
(13) ⟶ Auto fine tuning down
(14) ⟵ Remote control data
(15) ⟷ Keypad functions scanning
(16) ⟷ Keypad functions scanning
(17) ⟷ Keypad functions scanning
(18) ⟷ Keypad functions scanning
(19) ⟷ Keypad functions scanning
(20) Chassis
(21) Chassis
(22) ⟷ System oscillator 4 MHz
(23) ⟷ System oscillator 4 MHz
(24) ⟵ Power on reset = low
(25) ⟶ Power switch pulse contact

(Continued opposite)

(26) Chassis
(27) Chassis
(28) ⟶ TV /AV selection
(29) ⟶ Text control
(30) ⟵ Received signal ident pulse
(31) ⟷ Data external memory
(32) ⟶ On screen display
(33) ⟶ On screen display
(34) ⟶ External memory enable, data transfer = low
(35) ⟷ On screen display oscillator
(36) ⟶ Data, digital to analogue converter circuit & keypad functions scanning
(37) ⟷ Data, digital to analogue converter, external memory & keypad scanning
(38) ⟷ Data, digital to analogue converter, external memory & keypad scanning
(39) ⟷ Data, digital to analogue converter, external memory & keypad scanning
(40) ⟶ Clock, digital to analogue converter & external memory
(41) ⟵ 5 V supply
(42) ⟵ Vertical picture blanking pulse

Variation

Pins
(6), (7) N/c
(26) ⟶ 50 Hz/ 60 Hz scan system switch

Notes:

CXP80420-118S

CXP80420-126

Control microprocessor IC

Pins
(1) ——→ System clock and data disconnect
(2) N/c
(3) N/c
(4) N/c
(5) N/c
(6) N/c
(7) N/c
(8) N/c
(9) ←——→ Keypad functions scanning
(10) ←——→ Keypad functions scanning
(11) ←——→ Keypad functions scanning
(12) ←——→ Keypad functions scanning
(13) ——→ Text
(14) ——→ Change bleep sound to sound circuit
(15) ——→ Sound mute
(16) ——→ Picture mute
(17) N/c
(18) N/c
(19) N/c
(20) N/c
(21) ——→ Tuning band selection
(22) ——→ Tuning band selection
(23) ——→ Timer on LED display
(24) ——→ Timer off
(25) N/c
(26) N/c
(27) N/c
(28) ←—— Received signal ident pulse
(29) N/c
(30) N/c
(31) N/c
(32) Chassis
(33) N/c
(34) ←——→ System oscillator 4 MHz
(35) ←——→ System oscillator 4 MHz
(36) ←—— Power on reset = low

(Continued opposite)

(37)————→ Power on / standby control
(38)————→ Channel tuning control
(39)
(40)
(41)←———— Auto fine tuning
(42)←———— SCART socket 1 status pin 8
(43)←———— SCART socket 2 status pin 8
(44)←———— Remote control data
(45)←———→ On screen display oscillator
(46)←———→ On screen display oscillator
(47)←———— On screen display horizontal sync. pulse
(48)←———— On screen display vertical sync. pulse
(49)————→ Blue on screen display
(50)————→ Green on screen display
(51)————→ Red on screen display
(52)————→ On screen picture blanking
(53)←———→ System data bus
(54) N/c
(55)←———→ System clock bus
(56) N/c
(57) N/c
(58) N/c
(59) N/c
(60) N/c
(61)Chassis
(62)Chassis
(63) N/c
(64)←———— 5 V Supply

Variation

(21), (22), (24), (38) N/c

Notes:

CXP80420 - 127S

CXP80420 - 127S

Control microprocessor IC

Pins

(1) ←→ Keypad functions scanning
(2) ←→ Keypad functions scanning
(3) ←→ Keypad functions scanning
(4) ←→ Keypad functions scanning
(5) ←→ Keypad functions scanning
(6) ←→ Keypad functions scanning
(7) ←→ Keypad functions scanning
(8) ←→ Keypad functions scanning
(9) ←→ Keypad functions scanning
(10) ←→ Keypad functions scanning
(11) ←→ Keypad functions scanning
(12) ←→ Keypad functions scanning
(13) → Satellite tuning data
(14) → Change bleep to sound circuit
(15) → Sound mute
(16) → Picture/AV mute
(17) → Transmission system sound I.F select
(18) → Transmission system sound I.F select
(19) → Transmission system band select
(20) → TV I.F enable
(21) → Stereo LED display
(22) → Bilingual sound LED display
(23) → Timer mode on LED display
(24) N/c
(25) → TV/satellite enable
(26) → TV /satellite clock bus
(27) ←→ TV/satellite data bus
(28) ← Signal received ident pulse
(29) → NICAM decoder enable
(30) → NICAM mono 1 /mono 2 select
(31) → Stereo/mono select
(32) Chassis
(33) N/c
(34) ←→ System clock 4 MHz
(35) ←→ System clock 4 MHz
(36) ← Power on reset = Low
(37) → Power on/standby control

(Continued opposite)

(38) N/c
(39) ←——— Satellite sync signal
(40) ←——— SCART socket 2 status pin 8
(41) ←——— SCART socket 1 status pin 8
(42) ←——— Auto fine tuning
(43) ———→ Satellite mode enable
(44) ←——— Remote control data
(45) ←———→ On screen display oscillator
(46) ←———→ On screen display oscillator
(47) ←——— On screen display horizontal sync
(48) ←——— On screen display vertical sync
(49) ———→ Blue on screen display
(50) ———→ Green on screen display
(51) ———→ Red on screen display
(52) ———→ On screen display picture blanking
(53) ←———→ System data bus
(54) ———→ Sound effects woofer & LED display
(55) ←———→ System clock bus
(56) ←———→ Satellite LED display
(57) ←———→ Surround sound effect circuit enable
(58) ←———→ Surround sound effect balance control
(59) ←———→ Surround sound effect volume control
(60) ←———→ Sound volume control
(61) Chassis
(62) Chassis
(63) N/c
(64) ←——— 5 V supply

Variations

(17) to (22) N/c
(33) N/c
(57) to (60) N/c

Notes:

CX80424-107

CX80424-107

Control microprocessor IC

Pin
(1) ←——→ Keypad functions scanning
(2) ←——→ Keypad functions scanning
(3) ←——→ Keypad functions scanning
(4) ←——→ Keypad functions scanning
(5) ←——→ Keypad functions scanning
(6) ←——→ Keypad functions scanning
(7) ←——→ Keypad functions scanning
(8) ←——→ Keypad functions scanning
(9) N/c
(10) ←——→ Keypad functions scanning
(11) ←——→ Keypad functions scanning
(12) ←——→ Keypad functions scanning
(13) ——→ Teletext enable
(14) ——→ Change bleep to sound circuit
(15) ——→ Sound mute
(16) ——→ External sound/picture mute switching
(17) ——→ TV system sound I.F switching
(18) ——→ TV system sound I.F switching
(19) ——→ Multi TV system selection switching
(20) ——→ TV I.F switch
(21) ——→ Stereo LED display
(22) ——→ Bilingual LED display
(23) ——→ Timer mode LED display
(24) N/c
(25) N/c
(26) ←—— Remote control data
(27) N/c
(28) ←—— Received signal ident pulse
(29) ←—— NICAM decoder enable
(30) ——→ NICAM mono 1 /mono 2 select
(31) ——→ Stereo/mono select
(32) Chassis
(33) ——→ Auto tube degauss circuit
(34) ←——→ System clock oscillator 4 MHz
(35) ←——→ System clock oscillator 4 MHz
(36) ←—— Power on reset pulse = low

(Continued opposite)

(37) ⟶ Power on/standby control
(38) N/c
(39) ⟵ Super VHS input socket
(40) ⟵ SCART socket 2 status pin 8
(41) ⟵ SCART socket 1 status pin 8
(42) ⟵ Auto fine channel tuning
(43) ⟵ Vertical timebase pulse
(44) ⟵ As pin 26
(45) ⟷ On screen display oscillator
(46) ⟷ On screen display oscillator
(47) ⟵ On screen display horizontal sync pulse
(48) ⟵ On screen display vertical sync pulse
(49) ⟶ Blue on screen display
(50) ⟶ Green on screen display
(51) ⟶ Red on screen display
(52) ⟶ On screen display picture blanking
(53) ⟷ Serial data bus
(54) N/c
(55) ⟷ Serial clock bus
(56) ⟶ Dolby sound on/off switching
(57) ⟶ Surround sound on/off switching
(58) ⟶ Sound balance control
(59) ⟶ Surround sound volume control
(60) ⟶ Sound volume control
(61) Chassis
(62) Chassis
(63) N/c
(64) ⟵ 5 V supply

Notes:

CXP80424

Control microprocessor IC

Pins
(1) ———→ Tuner automatic gain control
(2) ———→ SCART socket 2 status pin 8
(3) ———→ SCART socket 1 status pin 8
(4) ———→ Power on/standby control
(5) ———→ Automatic degauss circuit
(6) N/c
(7) N/c
(8) ←——— Power rail failure sensing, settings memorized
(9) ———→ On screen display enable switching
(10) ———→ External red, green, blue picture blanking
(11) ———→ Teletext clock
(12) ←——— Beam current protection sensing
(13) N/c
(14) N/c
(15) N/c
(16) N/c
(17) N/c
(18) N/c
(19) ———→ Text circuit enable
(20) N/c
(21) N/c
(22)
(23)
(24) ←——— 5 V supply
(25)
(26) ←——— Remote control data
(27)
(28) ←——— Headphone inserted detection
(29) N/c
(30) ———→ Data & clock bus on =high, bus off & s/by indicator on = low
(31) ———→ Power on LED display = low
(32) Chassis
(33) N/c
(34) ←———→ System oscillator 4 MHz

(Continued opposite)

(35) ←→ System oscillator 4 MHz
(36) ← Power on reset = low
(37) N/c
(38) → Sound mute = low
(39) ←→ Keypad functions scanning
(40) ←→ Keypad functions scanning
(41) N/c
(42) → Auto fine tuning
(43) ← Received TV signal ident pulse
(44) ← Vertical timebase pulse
(45) ←→ On screen display oscillator
(46) ←→ On screen display oscillator
(47) ← On screen display horizontal sync pulse
(48) ← On screen display vertical sync pulse
(49) → Blue on screen display
(50) → Green on screen display
(51) → Red on screen display
(52) → On screen display picture blanking
(53) ←→ System data bus
(54) ←→ Teletext data bus
(55) → System clock bus
(56) → Teletext clock bus
(57) N/c
(58) N/c
(59) N/c
(60) → Picture sharpness control
(61) Chassis
(62) Chassis)
(63) ← 5 V supply
(64) ← 5 V supply

Notes:

CXP80424-143S

CXP80424-143S

Control microprocessor IC

Pins
(1) ————→ Clock/data bus disconnect
(2) N/c
(3) ←————→ Clock bus
(4) ←————→ Data bus
(5) ————→ Control bus
(6) N/c
(7) N/c
(8) ————→ External memory data transfer enable
(9) ←————→ Keypad functions scanning
(10) ←————→ Keypad functions scanning
(11) ←————→ Keypad functions scanning
(12) ←————→ Keypad functions scanning
(13) N/c
(14) ————→ Mode bleep to sound circuit
(15) ————→ Sound mute on = high
(16) ————→ TV/AV picture mute on = high
(17) ————→ Received sound transmission system PAL, SECAM select
(18) ————→ Received sound transmission system PAL, SECAM select
(19) ————→ Received sound transmission system PAL, SECAM select
(20) ————→ Received transmission system select PAL, SECAM
(21) N/c
(22) N/c
(23) N/c
(24) ————→ NICAM select
(25) ————→ Picture aspect select 4:3 = low, 16:9 = high
(26) N/c
(27) ←———— Received signal ident
(28) ←———— AV/teletext sync
(29) ←———— SCART socket 1 status pin 8
(30) ←———— SCART socket 2 status pin 8
(31) N/c
(32) Chassis
(33) ————→ Dolby sound on, surround sound off = low

(Continued opposite)

(34) ←——→ System clock oscillator 4 MHz
(35) ←——→ System clock oscillator 4 MHz
(36) ←—— Power on reset = Low
(37) ——→ Power on/standby control, standby = high
(38) N/c
(39) N/c
(40) N/c
(41) ←—— Auto fine tuning
(42) Chassis
(43) Chassis
(44) ←—— Remote control data
(45) ←——→ On screen display oscillator
(46) ←——→ On screen display oscillator
(47) ←—— On screen display horizontal sync pulse
(48) ←—— On screen display vertical sync pulse
(49) ——→ Blue on screen display
(50) ——→ Green on screen display
(51) ——→ Red on screen display
(52) ——→ On screen picture blanking
(53) ←——→ System clock bus
(54) ——→ Dolby sound balance control
(55) ←——→ System data bus
(56) N/c
(57) N/c
(58) N/c
(59) N/c
(60) N/c
(61) Chassis
(62) Chassis
(63) N/c
(64) ←—— 5 V supply

Variations

(17) to (20) N/c
(22) ←—— Received signal ident
(38) ——→ Channel tuning (tuner not clock/data bus driven)
(56) ——→ Sound effects echo volume control, karaoke system
(57) ——→ Microphone volume control, karaoke system
(58) ←——→ Karaoke system
(59) ←——→ Karaoke system
(60) ←——→ Karaoke system

CXP85116

CXP85116

Control microprocessor IC

Pins
(1) ⟶ Received system PAL, SECAM, NTSC select
(2) ⟶ Received system sound I.F select
(3) ⟶ TV sound mute
(4) ⟶ Sound I.F switching
(5)
(6)
(7)
(8) ⟶ Power on/standby control
(9) ⟵ SCART socket status pin 8
(10) ⟶ Change bleep to sound circuit
(11) Chassis
(12) ⟶ External RGB/TV blanking pulse switching
(13) ⟶ Sound mute
(14) N/c
(15) ⟶ External RGB on = low
(16) ⟶ External RGB on = high
(17) ⟶ TV /AV video switching
(18) ⟶ RGB tube drive mute
(19) ⟶ TV = high, AV = low
(20) ⟷ Keypad functions scanning
(21) ⟷ Keypad functions scanning
(23) ⟷ Keypad functions scanning
(24) ⟷ Keypad functions scanning
(25) ⟷ Keypad functions scanning
(26) ⟵ Remote control data
(27)
(28) ⟵ Received TV signal ident pulse
(29) ⟶ Standby LED display
(30)
(31)
(32) Chassis
(33) ⟶ Auto tube degauss circuit
(34) ⟷ System oscillator 4 MHz
(35) ⟷ System oscillator 4 MHz
(36) ⟵ Power on reset = low

(Continued opposite)

(37)——→ 50 Hz /60 Hz scan system switching
(38) N/c
(39) N/c
(40)←—— Power supply circuit
(41)←—— Power supply circuit
(42)←—— Auto fine tuning
(43)←——→ Clock bus external memory
(44) N/c
(45)←——→ On screen display oscillator
(46)←——→ On screen display oscillator
(47)←—— On screen display horizontal sync pulse
(48)←—— On screen display vertical sync pulse
(49)——→ Blue on screen display
(50)——→ Green on screen display
(51)——→ Red on screen display
(52)——→ On screen display picture blanking
(53)←——→ System data bus
(54)——→ NTSC picture hue control
(55)←——→ System clock bus
(56)——→ Picture contrast control
(57)——→ Picture sharpness control
(58)——→ Picture colour control
(59)——→ Picture brightness control
(60)——→ Sound volume control
(61) Chassis
(62) Chassis
(63)←—— 5 V supply
(64)←—— 5 V supply

Notes:

GS8014-04C

GS8014-04C

Control microprocessor IC

Pins
(1) ⟶ Tuning band select VHF-L
(2) ⟶ Tuning band select VHF-H
(3) ⟶ Tuning band select UHF
(4) N/c
(5) ⟷ Keypad function scanning
(6) ⟷ Keypad function scanning
(7) ⟷ Keypad function scanning
(8) ⟷ On screen display oscillator
(9) ⟷ On screen display oscillator
(10) ⟶ External circuits bus enable
(11) ⟷ I2C system data bus
(12) ⟶ I2C system clock bus
(13) ⟵ Received signal ident pulse
(14) ⟵ On screen display horizontal sync pulse
(15) ⟵ On screen display vertical sync pulse
(16) ⟶ On screen display picture blanking
(17) ⟶ Auto fine tuning mute
(18) ⟵ Auto fine tuning control
(19) Chassis
(20) ⟷ Keypad function scanning
(21) ⟷ Keypad function scanning
(22) ⟷ Keypad function scanning
(23) ⟷ Keypad function scanning
(24) Chassis
(25) ⟷ Keypad function scanning
(26) N/c
(27) N/c
(28)
(29) ⟵ Power on reset = low
(30) ⟷ System oscillator 4 MHz
(31) ⟷ System oscillator 4 MHz
(32) ⟶ TV/AV select
(33) N/c
(34) ⟶ Power on /standby control & LED display
(35) ⟶ On screen display picture blanking
(36)
(37) ⟶ Red on screen display

(Continued opposite)

(38) ⟶ Green on screen display
(39) ⟶ Blue on screen display
(40) N/c
(41) ⟶ TV channel tuning control
(42) ⟵ Remote control data
(43) ⟵ TV/ AV switching
(44) ⟶ Picture brightness control
(45) ⟶ Picture colour control
(46) ⟶ Picture contrast control
(47) ⟶ Sound volume control
(48) ⟵ 5 V supply

Notes:

HD401220

Control microprocessor IC

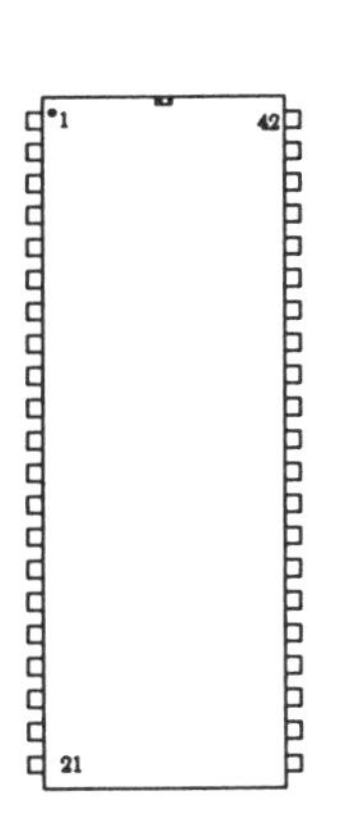

Pins
(1) Chassis
(2) ⟵ 5 V supply
(3) ⟷ Keypad functions scanning & 7 segment display decoder
(4) ⟷ Keypad functions scanning & 7 segment display decoder
(5) ⟷ Keypad functions scanning & 7 segment display decoder
(6) ⟷ Keypad functions scanning & 7 segment display decoder
(7) ⟷ Keypad functions scanning
(8) ⟷ Keypad functions scanning
(9) ⟷ Keypad functions scanning
(10) ⟶ 7 segment display
(11) ⟶ 7 segment display

(Continued overleaf)

HD401220

(12) ⟶ Tuning band selection
(13) ⟶ Tuning band selection
(14) ⟵ 5 V supply
(15) ⟶ Tuning band selection
(16) ⟶ Picture colour control
(17) ⟶ Picture contrast control
(18) ⟶ Picture brightness control
(19) ⟶ Sound volume control
(20) ⟵ Remote control data
(21) ⟶ Auto channel frequency control
(22) ⟶ Power on/standby control
(23) ⟶ 7 segment display decoder
(24) ⟵ Power on reset = low
(25) ⟶ Channel tuning control
(26) ⟵ 5 V supply
(27) ⟷ System oscillator 4 MHz
(28) ⟷ System oscillator 4 MHz

Variation

Pins
(10) N/c
(12) N/c
(13) N/c
(15) N/c

Notes:

HD401304

Control microprocessor IC

Pins

(1) Chassis
(2) ←——→ System oscillator 4 MHz
(3) ←——→ System oscillator 4 MHz
(4) Chassis
(5) Chassis
(6)
(7)
(8) ←——→ Keypad functions scanning
(9) ←——→ Keypad functions scanning
(10) ←——→ Keypad functions scanning
(11)
(12) ←—— Received signal ident pulse
(13)
(14) ←—— On screen display horizontal sync pulse
(15) ←—— On screen display vertical sync pulse
(16)
(17)
(18)
(19) ——→ On screen display character & picture blanking
(20)
(21) ←—— 5 V supply
(22) ——→ Tuning band select
(23) ——→ Tuning band select
(24) ←——→ Keypad functions scanning
(25) ←——→ Keypad functions scanning
(26)
(27)
(28) ←——→ Keypad functions scanning
(29) ←——→ Keypad functions scanning
(30) ——→ Power on/standby control
(31) ——→ Auto fine channel tuning
(32)
(33)
(34)
(35)
(36) ——→ Channel tuning control
(37) ←—— Auto fine channel tuning

Continued overleaf

(38)
(39) ← 50 Hz/60 Hz scanning system detect
(40) ← Remote control data
(41) ← Power on reset = low
(42) → Channel tuning control

Notes:

HD401304R12S

Control microprocessor IC

Pins
(1) Chassis
(2) ↔ System oscillator 4 MHz
(3) ↔ System oscillator 4 MHz
(4) Chassis
(5) Chassis
(6) ← SCART socket 1 status pin 8
(7) ← SCART socket 2 status pin 8
(8) N/c
(9) → high = surround sound effect
(10)
(11) ← Power switch pulse contact = high
(12)
(13) → Power on/standby control
(14) ← On screen display horizontal sync pulse
(15) ← On screen display vertical sync pulse
(16) N/c
(17) → Red on screen display
(18) → Green on screen display
(19) → On screen display picture blanking
(20) ↔ On screen display oscillator
(21) ← 5 V supply
(22) ↔ Keypad function scanning
(23) ↔ Keypad function scanning
(24) ↔ Keypad function scanning
(25) ↔ Keypad function scanning
(26) ↔ Keypad function scanning

(Continued opposite)

(27) ——→ TV/SCART 2 select (if Pin 7 high)
(28) ——→ TV = high, VTR = low
(29) ——→ Channel search = high, channel found = low
(30)
(31)
(32)
(33) ——→ Picture contrast control
(34) ——→ Picture colour control
(35) ——→ Picture brightness control
(36) ——→ Sound volume control
(37)
(38) ←—— Received signal ident pulse
(39)
(40) ←—— Remote control data
(41) ←—— Power on reset = low
(42) ——→ Channel tuning control

Notes:

HD401314-RA26S

Control microprocessor IC

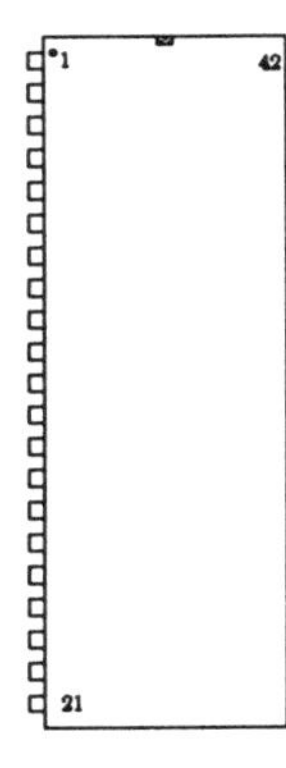

Pins
(1) Chassis
(2) ←——→ System oscillator 4 MHz
(3) N/c
(4) Chassis
(5) Chassis
(6) N/c
(7) N/c
(8) N/c
(9)
(10)
(11) ——→ Audio tone to sound circuit
(12) ——→ Timer mode on & LED display
(13) ——→ Power/standby control
(14) ←—— On screen display horizontal sync pulse
(15) ←—— On screen display vertical sync pulse

(Continued overleaf)

HD401314-RA26S

(16)——————→ Red on screen display
(17)——————→ Green on screen display
(18)——————→ Blue on screen display
(19)——————→ On screen display picture blanking
(20)←—————→ On screen oscillator
(21)←—————→ On screen oscillator
(22)←—————→ Keypad functions scanning
(23)←—————→ Keypad functions scanning
(24)←—————→ Keypad functions scanning
(25)←—————→ Keypad functions scanning
(26)←—————→ Keypad functions scanning
(27)——————→ AV LED display
(28)——————→ TV /AV select
(29)——————→ Auto fine channel tuning
(30) ←——————
(31) N/c
(32) N/c
(33)——————→ Picture contrast control
(34)——————→ Picture brightness control
(35)——————→ Picture colour control
(36)——————→ Sound volume control
(37)——————→ Auto frequency control
(38)←—————— Received signal ident pulse
(39)←—————— 50 Hz timer mode reference
(40)←—————— Remote control data
(41)←—————— Power on reset = low
(42)——————→ Channel tuning control

Variation

(19) N/c

....

Notes:

IRT1250

Infra-red remote controller IC

Pins
(1) Ground
(2) ↔ RC components (system oscillator)
(3) ↔ RC components (system oscillator)
(4) ↔ RC components (system oscillator)
(5) ← Infra-red transmitting diode driver
(6) ← 9 V supply
(7) ← 9 V supply
(8) ↔ Keypad functions scanning
(9) ↔ Keypad functions scanning
(10) ↔ Keypad functions scanning
(11) ↔ Keypad functions scanning
(12) ↔ Keypad functions scanning
(13) ↔ Keypad functions scanning
(14) ↔ Keypad functions scanning
(15) ↔ Keypad functions scanning
(16) ↔ Keypad functions scanning
(17) ↔ Keypad functions scanning
(18) ↔ Keypad functions scanning
(19) ↔ Keypad functions scanning
(20) ↔ Keypad functions scanning
(21) ↔ Keypad functions scanning
(22) ↔ Keypad functions scanning
(23) ↔ Keypad functions scanning

Notes:

LC7462M

Infra-red remote controller IC

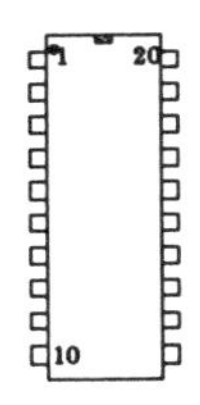

Pins
(1) ↔ Keypad functions scanning
(2) ↔ Keypad functions scanning
(3) ↔ Keypad functions scanning

(Continued overleaf)

M708L

(4) ←——→ Keypad functions scanning
(5) ——→ Infra-red transmitting diode driver
(6) ←—— 3 V supply
(7) ←—— 3 V supply
(8) ←——→ System oscillator 455 kHz
(9) ←——→ System oscillator 455 kHz
(10) Ground
(11) N/c
(12) ←——→ Keypad functions scanning
(13) ←——→ Keypad functions scanning
(14) ←——→ Keypad functions scanning
(15) ←——→ Keypad functions scanning
(16) ←——→ Keypad functions scanning
(17) ←——→ Keypad functions scanning
(18) ←——→ Keypad functions scanning
(19) ←——→ Keypad functions scanning
(20) ←—— 3 V supply

Notes:

M708L

Infra-red remote controller IC

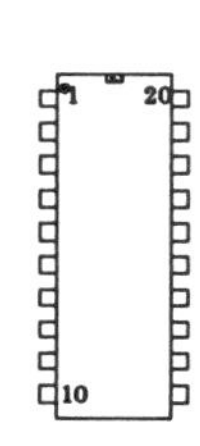

Pins
(1) ←—— 3 V supply
(2) ←——→ Oscillator 455 kHz
(3) ←——→ Oscillator 455 kHz
(4) ←——→ Keypad function scanning
(5) ←——→ Keypad function scanning
(6) ←——→ Keypad function scanning
(7) ←——→ Keypad function scanning
(8) ←——→ Keypad function scanning
(9) ←——→ Keypad function scanning
(10) ←——→ Keypad function scanning
(11) ←——→ Keypad function scanning
(12) ←——→ Keypad function scanning
(13) ←——→ Keypad function scanning
(14) ←——→ Keypad function scanning

(Continued opposite)

(15) Chassis
(16) ←——— 3 V supply
(17) Chassis
(18) Chassis
(19) ———→ Data to infra-red transmitter diode driver
(20) ←——— 3 V supply

Notes:

M710B1

Infra-red remote controller IC

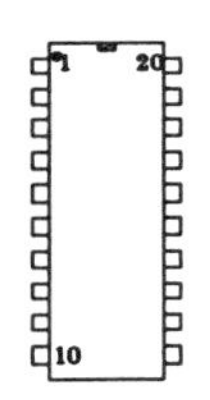

Pins
(1) Ground
(2) ←———→ System oscillator 500 kHz
(3) ←———→ System oscillator 500 kHz
(4) ←———→ Keypad functions scanning
(5) ←———→ Keypad functions scanning
(6) ←———→ Keypad functions scanning
(7) ←———→ Keypad functions scanning
(8) ←———→ Keypad functions scanning
(9) ←———→ Keypad functions scanning
(10) ←———→ Keypad functions scanning
(11) ←———→ Keypad functions scanning
(12) ←———→ Keypad functions scanning
(13) ←———→ Keypad functions scanning
(14) ←———→ Keypad functions scanning
(15) N/c
(16) ←———→ Keypad functions scanning
(17) ←———→ Keypad functions scanning
(18) ←———→ Keypad functions scanning
(19) ←———→ Keypad functions scanning
(20) ←———→ Keypad functions scanning
(21) Ground
(22) Ground
(23) ←——— 9 V supply
(24) Ground
(25) Ground

(Continued overleaf)

(26) Ground
(27) ⟶ Infra-red transmitting diode driver
(28) ⟵ 9 V supply

Notes:

M3004LABI

Infra-red remote controller IC

Pins
(1) ⟶ Infra-red transmitting diode driver
(2) ⟷ Keypad functions scanning
(3) ⟷ Keypad functions scanning
(4) N/c
(5) ⟷ Keypad functions scanning
(6) ⟷ Keypad functions scanning
(7) ⟷ Keypad functions scanning
(8) ⟷ Keypad functions scanning
(9) ⟷ Keypad functions scanning
(10) Ground
(11) ⟷ System oscillator 400 kHz
(12) ⟷ System oscillator 400 kHz
(13) ⟷ Keypad functions scanning

(Continued overleaf)

(14) ⟷ Keypad functions scanning
(15) ⟷ Keypad functions scanning
(16) ⟷ Keypad functions scanning
(17) N/c
(18) ⟷ Keypad functions scanning
(19) ⟷ Keypad functions scanning
(20) ⟵ 3 V supply

Notes:

M3005

Infra-red remote controller IC

Pins
(1) ⟶ Infra-red transmitting diode driver
(2) ⟷ Keypad functions scanning
(3) ⟷ Keypad functions scanning
(4) ⟷ Keypad functions scanning
(5) ⟷ Keypad functions scanning
(6) ⟷ Keypad functions scanning
(7) ⟷ Keypad functions scanning
(8) ⟷ Keypad functions scanning
(9) ⟷ Keypad functions scanning
(10) Ground
(11) ⟷ System oscillator 455 kHz
(12) ⟷ System oscillator 455 kHz
(13) ⟷ Keypad functions scanning
(14) ⟷ Keypad functions scanning
(15) ⟷ Keypad functions scanning
(16) ⟷ Keypad functions scanning
(17) ⟷ Keypad functions scanning
(18) ⟷ Keypad functions scanning
(19) ⟷ Keypad functions scanning
(20) ⟵ 3 V supply

Notes:

M491BB1

M491BB1

Control microprocessor IC

Pins
(1) Chassis
(2) ←——— Read/write/erase memory mode control
(3) ———→ Data depending on key-scan operation to pin 2
(4) ———→ Auto channel fine tuning control
(5) ———→ Channel tuning control
(6) ———→ Auto channel fine tuning control
(7) ←——→ System oscillator 455 kHz
(8) ←——→ System oscillator 455 kHz
(9) ←——— 5 V supply
(10) Chassis
(11) ←——— Remote control data
(12)
(13)
(14) ←——→ 7 segment LED display
(15) ———→ Sound volume control & sound mute
(16) ←——→ Keypad function scanning
(17) Chassis
(18) ←——→ Keypad function scanning
(19) ←——→ Keypad function scanning
(20) ←——→ Keypad function scanning
(21) ←——→ Keypad function scanning
(22) ←——→ Keypad function scanning
(23) ←——→ Keypad function scanning
(24) ←——→ Keypad function scanning
(25) ←——— Power on reset = low
(26) ———→ Power on/standby control (standby = low)
(27) ———→ 7 segment LED display
(28) ———→ 7 segment LED display
(29) ———→ 7 segment LED display
(30) ———→ 7 segment LED display
(31) ———→ 7 segment LED display
(32) ———→ 7 segment LED display
(33) ———→ 7 segment LED display
(34) ———→ 7 segment LED display
(35) ———→ 7 segment LED display
(36) ———→ 7 segment LED display

(Continued opposite)

1 40 20

(37) ——→ Tuning band select
(38) Chassis
(39) Chassis
(40) Chassis

Variation

Pins
(4), (5), (14) N/c

Notes:

M50120P

Control microprocessor IC

Pins
(1) Chassis
(2) ←—— Remote control data
(3) ←—— Power on reset = low
(4)
(5)
(6) ——→ Sound volume increase function switch
(7) ——→ Sound volume decrease function switch
(8)
(9) ←—— Sound volume function switches
(10)
(11)
(12) ←——→ System oscillator 455 kHz
(13) ←——→ System oscillator 455 kHz
(14) ←—— Power on/off switch pulse contact = high
(15) ←—— 12 V supply
(16) ——→ Picture colour control
(17) ——→ Picture brightness control
(18) ——→ Picture contrast control
(19) ——→ Sound volume control
(20) N/c
(21) N/c
(22) ←——→ System data

(Continued overleaf)

M50431-101

(23) ←——→ Teletext data limited clock control
(24) N/c
(25) Chassis
(26) N/c
(27) ——→ Channel tuning up
(28) ——→ Channel tuning down
(29)
(30) N/c

Variation

Pins
(6), (7), (9) N/c

Notes:

M50431-101

Control microprocessor IC

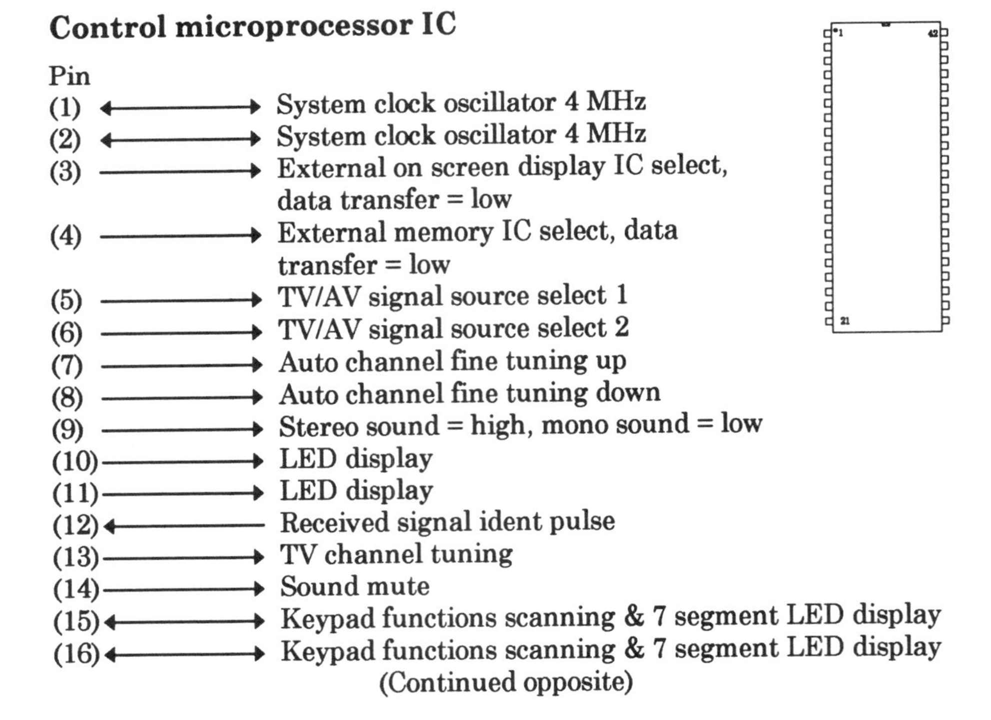

Pin
(1) ←——→ System clock oscillator 4 MHz
(2) ←——→ System clock oscillator 4 MHz
(3) ——→ External on screen display IC select, data transfer = low
(4) ——→ External memory IC select, data transfer = low
(5) ——→ TV/AV signal source select 1
(6) ——→ TV/AV signal source select 2
(7) ——→ Auto channel fine tuning up
(8) ——→ Auto channel fine tuning down
(9) ——→ Stereo sound = high, mono sound = low
(10) ——→ LED display
(11) ——→ LED display
(12) ←—— Received signal ident pulse
(13) ——→ TV channel tuning
(14) ——→ Sound mute
(15) ←——→ Keypad functions scanning & 7 segment LED display
(16) ←——→ Keypad functions scanning & 7 segment LED display

(Continued opposite)

(17)←——→ Keypad functions scanning & 7 segment LED display
(18)←——→ Keypad functions scanning & 7 segment LED display
(19)←——→ Keypad functions scanning & 7 segment LED display
(20)←——→ Keypad functions scanning & 7 segment LED display
(21)←——→ Keypad functions scanning & 7 segment LED display
(22) Chassis
(23) Chassis
(24)——→ Tuning band select, UHF = low, VHF = high
(25)——→ Tuning band select, UHF = high, VHF = low
(26) N/c
(27)——→ Power on/standby control
(28)——→ Sound volume control
(29)——→ Picture brightness control
(30)——→ Picture colour control
(31)——→ External memory IC mode control
(32)——→ External memory IC mode control
(33)——→ External memory IC mode control
(34)——→ Clock bus external memory & on screen display ICs
(35)←—— Remote control data
(36)←—— Power on reset = low
(37)←——→ Data bus external memory & on screen display ICs
(38)←——→ Keypad functions scanning
(39)←——→ Keypad functions scanning
(40)←——→ Keypad functions scanning
(41)←——→ Keypad functions scanning
(42)←—— 5 V supply

Variations

(5), (9), (10), (11) N/c
(14), (24), (25) N/c

Notes:

M50431 - 512SP

M50431 - 512SP

Control microprocessor IC

Pins

(1) ←——→ System oscillator 4 MHz
(2) ←——→ System oscillator 4 MHz
(3) ——→ Keypad function scanning & external memory IC clock
(4) ——→ External memory IC select, data transfer = low
(5) ——→ Keypad function scanning & external memory IC mode control
(6) ——→ Keypad function scanning & external memory IC mode control
(7) ——→ Keypad function scanning & external memory IC mode control
(8) ←——→ External memory IC data
(9) N/c
(10)——→ Auto channel fine tuning down control
(11)——→ Auto channel fine tuning up control
(12)←—— Received channel ident pulse
(13)——→ TV channel tuning control
(14)——→ Seven segment LED display
(15)——→ Seven segment LED display
(16)——→ Seven segment LED display
(17)——→ Seven segment LED display
(18)——→ Seven segment LED display
(19)——→ Seven segment LED display
(20)——→ Seven segment LED display
(21)——→ Seven segment LED display
(22) Chassis
(23) Chassis
(24) N/c
(25) N/c
(26)——→ Auto tuner & I.F gain control
(27)——→ Power on/standby control
(28)——→ Sound volume control
(29)——→ Picture contrast control
(30)——→ Picture colour control
(31)——→ Vertical blanking
(32) N/c

(Continued opposite)

(33) N/c
(34) N/c
(35) ←——— Remote control data
(36) ←——— Power on reset = low
(37) ———→ Sound mute
(38) ←———→ Keypad function scanning
(39) ←———→ Keypad function scanning
(40) ←———→ Keypad function scanning
(41) ←———→ Keypad function scanning
(42) ←——— 5 V supply

Notes:

M50431-563SP

Control microprocessor IC

Pins
(1) ←———→ System clock oscillator 4 MHz
(2) ←———→ System clock oscillator 4 MHz
(3) N/c
(4) ———→ External on screen display IC select, data transfer = low
(5)
(6) ←———→ Keypad function scanning
(7) ←———→ Keypad function scanning
(8) ———→ External memory IC select, data transfer = low
(9)
(10)
(11)
(12) ←——— Received signal ident pulse
(13) ———→ Channel tuning
(14) ———→ Muting
(15) ———→ RGB switching
(16) ———→ AV2 source switching
(17) ———→ AV1 source switching
(18)
(19)

(Continued overleaf)

(20)
(21)
(22) Chassis
(23) Chassis
(24) ⟶ Channel tuning band select
(25) ⟶ Channel tuning band select
(26) ⟶ Auto fine tuning
(27) ⟶ Power on/standby control
(28) ⟶ Sound volume control
(29) ⟶ Sound balance control
(30) ⟶ External on screen display IC clock
(31) ⟶ External memory IC clock
(32) ⟶ External memory IC mode control
(33) ⟶ External memory IC mode control
(34) ⟶ External memory IC mode control
(35) ⟵ Remote control data
(36) ⟵ Power on reset = low
(37) ⟷ External memory IC data
(38) ⟷ Keypad function scanning
(39) ⟷ Keypad function scanning
(40) ⟷ Keypad function scanning
(41)
(42) ⟵ 5 V supply

Variation

Pins
(12) N/c
(14) to (21) Fixed low

Notes:

M50433 - 531SP

Control microprocessor IC

Pins
(1) ←——— 5 V supply
(2) ———→ TV channel tuning
(3) ———→ Red on screen display
(4) ———→ Green on screen display
(5) ———→ Blue on screen display
(6) ———→ On screen display picture blanking
(7) ←——→ On screen display oscillator
(8) ←——→ On screen display oscillator
(9) ←——— On screen display vertical sync pulse
(10) ←——— On screen display horizontal sync pulse
(11) ←——— Power on reset = low
(12) ←——— Remote control data
(13) ———→ TV/AV select
(14) N/c
(15) ———→ Sound mute
(16) ———→ Power on/standby control
(17) N/c
(18) N/c
(19) ———→ Timer function LED display
(20) ———→ External memory IC select, data transfer = low
(21) Chassis
(22) Chassis
(23) ———→ TV tuning band select
(24) ———→ TV tuning band select
(25) ———→ TV tuning band select
(26) ———→ TV channel auto fine tuning
(27) ———→ Sound volume control
(28) ←——→ Keypad functions scanning & external memory IC mode control
(29) ←——→ Keypad functions scanning & external memory IC mode control
(30) ←——→ Keypad functions scanning & external memory IC mode control
(31) ———→ External memory clock
(32) ←——→ External memory IC data
(33) ←——→ System clock oscillator
(34) ←——→ System clock oscillator

(Continued overleaf)

(35) Chassis
(36) ←———— 50 Hz/60 Hz scanning system detect
(37) ←————→ Keypad functions scanning
(38) ←————→ Keypad functions scanning
(39) ←————→ Keypad functions scanning
(40) ←————→ Keypad functions scanning
(41) Chassis
(42) ←———— Received signal ident pulse

Notes:

M50433B-502SP

Control microprocessor IC

Pins
(1) ←———— 5 V supply
(2) ————→ Channel tuning
(3) ————→ Red on screen display
(4) ————→ Green on screen display
(5) ————→ Blue on screen display
(6) ————→ On screen display picture blanking
(7) ←————→ On screen display oscillator
(8) ←————→ On screen display oscillator
(9) ←———— On screen display horizontal sync pulse
(10) ←———— On screen display vertical sync pulse
(11) ←———— Power on reset = low
(12) ←———— Remote control data
(13) N/c
(14) N/c
(15) ————→ TV/AV select
(16) ————→ Power on/standby control
(17) ————→ Auto fine tuning control
(18)
(19)
(20)
(21) Chassis
(22) Chassis
(23) N/c

(Continued opposite)

(24) N/c
(25) N/c
(26) ⟶ External memory
(27) ⟶ Sound mute
(28) ⟶ External memory mode control & keypad function scanning
(29) ⟶ External memory mode control & keypad function scanning
(30) ⟶ external memory mode control & keypad function scanning
(31) ⟶ External memory clock
(32) ⟷ External memory data
(33) ⟶ System clock oscillator
(34) ⟶ System clock oscillator
(35) ⟷ Keypad function scanning
(36) Chassis
(37) ⟷ Keypad function scanning
(38) ⟷ Keypad function scanning
(39) ⟷ Keypad function scanning
(40) ⟷ Keypad function scanning
(41)
(42)

Variation

Pins
(3), (4), (5) N/c

Notes:

M50435 - 893FP

M50435 - 893FP

Control microprocessor IC

Pins
(1) ←——— 5 V supply
(2) ———→ TV channel tuning
(3) ———→ Red on screen display
(4) ———→ Green on screen display
(5) ———→ Blue on screen display
(6) ———→ On screen display picture blanking
(7) ←——→ On screen display oscillator
(8) ←——→ On screen display oscillator
(9) ←——— On screen display vertical sync pulse
(10) ←——— On screen display horizontal sync pulse
(11)
(12) ←——— Remote control data
(13) ———→ Power switch pulse contact
(14) ———→ Power on/standby control
(15) ———→ TV/AV signal source select
(16) ———→ TV /AV signal source select
(17) ←——→ Keypad functions scanning
(18) ←——→ Keypad functions scanning
(19) ———→ External memory IC select, data transfer = low
(20) ———→ External digital to analogue converter IC data
(21) Chassis
(22) Chassis
(23) ———→ TV tuning band select, UHF
(24) ———→ TV tuning band select, VHF H
(25) ———→ TV tuning band select, VHF L
(26) ———→ Sound mute
(27) ———→ Sound volume control
(28) ←——→ Keypad functions scanning & external memory IC mode control
(29) ←——→ Keypad functions scanning & external memory IC mode control
(30) ←——→ Keypad functions scanning & external memory IC mode control
(31) ———→ External memory IC clock
(32) ←——→ External memory IC data
(33) ←——→ System clock oscillator
(34) ←——→ System clock oscillator

(Continued opposite)

(35)⟶ Timer function on
(36)⟶ Timer function off
(37)⟷ Keypad functions scanning
(38)⟷ Keypad functions scanning
(39)⟷ Keypad functions scanning
(40)⟷ Keypad functions scanning
(41)⟷ Keypad functions scanning
(42)⟶ Transmission PAL, SECAM, NTSC system select

Notes:

M50436 - 587SP

Control microprocessor IC

Pins
(1) ⟶ TV channel tuning control
(2) N/c
(3) ⟶ Standby LED indicator
(4) N/c
(5) ⟵ Remote control data
(6) ⟶ Power on = low/standby = high control
(7) ⟶ Tuning band select, UHF = high, VHF H = low, VHF L = low
(8) ⟶ Tuning band select, UHF = low, VHF H = high, VHF L = low
(9) ⟶ NTSC 3.58 MHz reset to external sub processor = low
(10)⟶ NTSC 4.43 MHz enable signal to external sub-processor = low
(11)⟶ SECAM/PAL switching
(12)⟶ SECAM/PAL switching
(13)⟵ 50 Hz/60 Hz scan system data, external sub processor
(14)⟵ NTSC transmission system data, external sub-processor
(15)⟵ SECAM transmission system data, external sub-processor
(16)⟵ PAL transmission system data, external sub-processor
(17)⟷ Keypad function scanning
(18)⟷ Keypad function scanning

(Continued overleaf)

(19) ←——→ Keypad function scanning
(20) ←——→ Keypad function scanning
(21) ←——→ Keypad function scanning
(22) ←——→ Keypad function scanning
(23) ←——→ Keypad function scanning
(24) ——→ Control signal to external digital to analogue converter IC
(25) Chassis
(26) Chassis
(27) ←—— Power on reset = low
(28) ←——→ System clock 4 MHz
(29) ←——→ System clock 4 MHz
(30) ←——→ Keypad function scanning & external memory clock
(31) ←——→ Keypad function scanning
(32) ←——→ Keypad function scanning
(33) ←——→ Keypad function scanning
(34) ←——→ Data, external memory & digital to analogue converter ICs
(35) ←—— Channel auto fine tuning control
(36) ←—— Received signal ident pulse
(37) ←—— TV = high, AV = low (SCART status pin 8 = high)
(38) ←—— TV = high, AV = low (SCART status pin 8 = low)
(39) ←—— Additional sound/picture muting
(40) ——→ External memory IC select, data transfer = low
(41) ——→ Sound mute
(42)
(43) ——→ Auto channel search
(44) ——→ On screen display picture blanking
(45) ——→ Blue on screen display
(46) ——→ Green on screen display
(47) ——→ Red on screen display
(48) ←——→ On screen display oscillator
(49) ←——→ On screen display oscillator
(50) ←—— On screen display vertical sync pulse
(51) ←—— On screen display horizontal sync pulse
(52) ←—— 5 V supply

Notes:

M50436-588SP

Control microprocessor IC

Pins
(1) ⟶ TV channel tuning
(2) N/c
(3) N/c
(4) N/c
(5) ⟵ Remote control data
(6) ⟶ Power on/standby control
(7) ⟶ Tuning band VHF/UHF select
(8) ⟶ Tuning band VHF/UHF select
(9) N/c
(10) N/c
(11) ⟶ PAL/SECAM transmission select
(12) ⟶ PAL/SECAM transmission select
(13) N/c
(14) N/c
(15) N/c
(16) N/c
(17) ⟷ Keypad function scanning
(18) ⟷ Keypad function scanning
(19) ⟷ Keypad function scanning
(20) ⟷ Keypad function scanning
(21) N/c
(22) ⟷ Keypad function scanning
(23) N/c
(24) ⟶ Clock external digital to analogue converter IC
(25) Chassis
(27) Chassis
(28) ⟷ System oscillator 4 MHz
(29) ⟷ System oscillator 4 MHz
(30) ⟶ External memory & digital to analogue converter ICs clock
(31) ⟷ Keypad function scanning & external memory IC mode control
(32) ⟷ Keypad function scanning & external memory IC mode control
(33) ⟷ Keypad function scanning & external memory IC mode control
(34) ⟷ Data external memory & digital to analogue ICs

(Continued overleaf)

M50436 - 614SP

(35) ⟶ TV auto channel fine tuning
(36) ⟵ Received signal ident pulse
(37)
(38)
(39)
(40) ⟶ External memory IC select, data transfer = low
(41) ⟶ Sound mute
(42) ⟶ Picture muting
(43) N/c
(44) ⟶ On screen display picture blanking
(45) ⟶ Blue on screen display
(46) ⟶ Green on screen display
(47) ⟶ Red on screen display
(48) ⟷ On screen display oscillator
(49) ⟷ On screen display oscillator
(50) ⟵ On screen display vertical sync pulse
(51) ⟵ On screen display horizontal sync pulse
(52) ⟵ 5 V supply

Notes:

M50436 - 614SP

Control microprocessor IC

Pins
(1) ⟶ TV channel tuning control
(2) ⟶ Sound volume control
(3) N/c
(4) N/c
(5) ⟵ Remote control data
(6) ⟶ Auto channel frequency control
(7) N/c
(8) ⟶ External memory IC select, data transfer = low
(9) ⟷ Teletext data – external processor
(10) ⟷ Data bus – external processor
(11) ⟷ Data bus – external processor

(Continued opposite)

(12) ←——→ Data bus – external processor
(13) ——→ Clock bus – external processor
(14) ←——→ Data request – external processor
(15) ←——→ Teletext data – external processor
(16) ——→ Power on/standby control
(17) ←—— 50 Hz/60 Hz scan system switching
(18) ←——→ Keypad function scanning
(19) ←——→ Keypad function scanning
(20) ←——→ Keypad function scanning
(21) ——→ Sound mute
(22) ——→ Auto tuner gain control
(23) ——→ Auto channel fine tuning control down
(24) ——→ Standby LED indicator
(25) Chassis
(26) Chassis
(27) ←—— Power on reset = low
(28) ←——→ System oscillator 4 MHz
(29) ←——→ System oscillator 4 MHz
(30) ←——→ Keypad function scanning & external memory IC clock
(31) ←——→ Keypad function scanning & external memory IC mode control
(32) ←——→ Keypad function scanning & external memory IC mode control
(33) ←——→ Keypad function scanning & external memory IC mode control
(34) ←——→ External memory IC data
(35) ——→ Auto channel fine tuning control up
(36) ←—— Received signal ident pulse
(37) ←—— SCART socket status pin 8
(38) N/c
(39) Chassis
(40) ——→ Tuning band select UHF
(41) ——→ Tuning band select VSH
(42) ——→ Tuning band select VHF H
(43) ——→ Tuning band select VHF L
(44) N/c
(45) ——→ Blue on screen display
(46) ——→ Green on screen display
(47) ——→ Red on screen display
(48) ←——→ On screen display oscillator
(49) ←——→ On screen display oscillator

(Continued overleaf)

M50436 - 616SP

(50) ⟶ On screen display vertical sync pulse
(51) ⟶ On screen display horizontal sync pulse
(52) ⟵ 5 V supply

Notes:

M50436 - 616SP

M50436 - 511SP

M50436 - 515SP

Control microprocessor IC

Pins
(1) ⟶ TV channel tuning
(2) ⟶ Sound volume control
(3) ⟶ Picture contrast control
(4) ⟶ Picture colour control
(5) ⟵ Remote control data
(6) ⟶ Auto channel frequency control
(7) ⟶ Sound I.F mute
(8) ⟶ External memory IC select, data transfer = low
(9) N/c
(10) ⟶ Picture blanking
(11) ⟶ Power on/standby control
(12) ⟶ Auto tuner gain control
(13) ⟵ SCART status socket 1 pin 8
(14) ⟵ SCART status socket 2 pin 8
(15) ⟶ AV1/AV2 select
(16) ⟶ TV = high, AV = low
(17) ⟵ 50 Hz/60 Hz scan system detect
(18) ⟷ Keypad function scanning
(19) ⟷ Keypad function scanning
(20) ⟷ Keypad function scanning
(21) N/c
(22) ⟶ Sound loudness control

(Continued opposite)

(23) ——→ Auto channel fine tuning control 1
(24) ——→ Standby LED indicator
(25) Chassis
(26) Chassis
(27) ←—— Power on reset = low
(28) ←——→ System clock oscillator 4 MHz
(29) ←——→ System clock oscillator 4 MHz
(30) ←——→ Keypad function scanning & external memory IC clock bus
(31) ←——→ Keypad function scanning & external memory IC mode control
(32) ←——→ Keypad function scanning & external memory IC mode control
(33) ←——→ Keypad function scanning & external memory IC mode control
(34) ←——→ Data bus
(35) ——→ Auto channel fine tuning 2
(36) ←—— Received signal ident pulse
(37) ←——→ Teletext data
(38) Chassis
(39) Chassis
(40) ——→ Tuning band select UHF/ VHF H/ VHF L
(41) ——→ Tuning band select UHF/ VHF H/ VHF L
(42) ——→ Tuning band select UHF/ VHF H/ VHF L
(43) ——→ Tuning band select UHF/ VHF H/ VHF L
(44) ——→ On screen display picture blanking
(45) ——→ Blue on screen display
(46) ——→ Green on screen display
(47) ——→ Red on screen display
(48) ←——→ On screen display oscillator
(49) ←——→ On screen display oscillator
(50) ←—— On screen display vertical sync pulse
(51) ←—— On screen display horizontal sync pulse
(52) ←—— 5 V supply

Variation

Pins
(13), (15), (17) Chassis

Notes:

M50436 - 589SP

Control microprocessor IC

Pins

(1) ——→ TV channel tuning control
(2) ——→ Picture colour control
(3) ——→ Picture brightness control
(4) ——→ Sound volume control
(5) ←—— Remote control data
(6) ——→ Power / standby LED indicator
(7) ——→ Tuning band selection
(8) ——→ Tuning band selection
(9) ——→ Transmission system NTSC / SECAM / PAL switching
(10) ——→ Transmission system NTSC / SECAM / PAL switching
(11) ——→ Transmission system NTSC / SECAM / PAL switching
(12) ——→ Transmission system NTSC / SECAM / PAL switching
(13) ←—— 50 Hz / 60 Hz scan system detect
(14) ←—— NTSC transmission switch
(15) ←—— SECAM transmission switch
(16) ←—— PAL transmission switch
(17) ←——→ Keypad function scanning
(18) ←——→ Keypad function scanning
(19) ←——→ Keypad function scanning
(20) ←——→ Keypad function scanning
(21) ←——→ Keypad function scanning
(22) ←——→ Keypad function scanning
(23) ←——→ Keypad function scanning
(24) N/c
(25) Chassis
(26) Chassis
(27) ←—— Power on reset = Low
(28) ←——→ System clock 4 MHz
(29) ←——→ System clock 4 MHz
(30) ——→ Keypad function scanning, external memory IC & system clock
(31) ——→ External memory mode control
(32) ——→ External memory mode control
(33) ——→ External memory mode control
(34) ←——→ Data bus & external memory IC

(Continued opposite)

(35)———→ Auto channel fine tuning
(36)←——— Received signal ident pulse
(37) N/c
(38) N/c
(39) N/c
(40)———→ External memory IC select, data transfer = low
(41)———→ Sound mute
(42)———→ TV picture mute
(43)
(44)———→ On screen display picture blanking
(45)———→ Blue on screen display
(46)———→ Green on screen display
(47)———→ Red on screen display
(48)←——→ On screen display oscillator
(49)←——→ On screen display oscillator
(50)←——— On screen display vertical sync. pulse
(51)←——— On screen display horizontal sync. pulse
(52)←——— 5v supply

Notes:

M50439 - 563P

Control microprocessor IC

Pins
(1) ———→ Sound volume control
(2) ———→ Picture colour control
(3) ———→ Picture contrast control
(4) ———→ Picture brightness control
(5) ←——— Remote control data
(6) ———→ External memory IC mode control
(7) ———→ External memory IC mode control
(8) ———→ External memory IC mode control
(9) ———→ TV channel tuning
(10)———→ Power on/standby control
(11)←——→ Keypad functions scanning
(12)←——→ Keypad functions scanning

(Continued overleaf)

(13)←——→ Keypad functions scanning
(14)←——→ Keypad functions scanning
(15)←——→ Keypad functions scanning
(16)←——→ Keypad functions scanning
(17)←——→ Keypad functions scanning
(18)←——→ Keypad functions scanning
(19)←——→ Keypad functions scanning
(20)←——→ Keypad functions scanning
(21)——→ External memory IC select, data transfer = low
(22)——→ TV tuning band select High
(23)——→ TV tuning band select low
(24)——→ TV channel auto fine tuning correction
(25) Chassis
(26) Chassis
(27)←—— Power on reset = low
(28)←——→ System oscillator 4 MHz
(29)←——→ System oscillator 4 MHz
(30)
(31)←—— SCART socket status pin 8
(32)
(33)
(34)←——→ External memory IC data
(35)←—— TV channel auto fine tuning
(36)←—— Received signal ident pulse
(37)——→ Teletext clock bus
(38)←——→ Teletext data bus
(39)——→ Picture blanking
(40)
(41)
(42)
(43)
(44)——→ On screen display picture blanking
(45)——→ Red on screen display
(46)——→ Green on screen display
(47)——→ Blue on screen display
(48)←——→ On screen display oscillator
(49)←——→ On screen display oscillator
(50)——→ On screen display vertical sync pulse
(51)——→ On screen display horizontal sync pulse
(52)←—— 5 V supply

Notes:

M50460-001P

M50560-001P

Infra-red remote controller IC

Pins
(1) Ground
(2) Ground
(3)
(4) ←——→ System oscillator 455 kHz
(5) ←——→ System oscillator 455 kHz
(6) ←——→ Keypad functions scanning
(7) ←——→ Keypad functions scanning
(8) ←——→ Keypad functions scanning
(9) ←——→ Keypad functions scanning
(10) ←——→ Keypad functions scanning
(11) N/c
(12) ←——→ Keypad functions scanning
(13) ←——→ Keypad functions scanning
(14) ←——→ Keypad functions scanning
(15) ←——→ Keypad functions scanning
(16) ←——→ Keypad functions scanning
(18) ←——→ Keypad functions scanning
(19) ——→ Infra-red transmitting diode driver
(20) ←—— 3 V supply

Notes:

M50460-033

M50460-033

M50560-033

Infra-red remote controller IC

Pins
(1) Ground
(2) Ground
(3) N/c
(4) ←——→ System oscillator 455 kHz
(5) ←——→ System oscillator 455 kHz
(6) N/c
(7) N/c
(8) ←——→ Keypad functions scanning
(9) N/c
(10) N/c
(11) N/c
(12) ←——→ Keypad functions scanning
(13) ←——→ Keypad functions scanning
(14) ←——→ Keypad functions scanning
(15) ←——→ Keypad functions scanning
(16) N/c
(17) N/c
(18) ←——→ Keypad functions scanning
(19) ←——→ Keypad functions scanning
(20) ←——→ Keypad functions scanning
(21) ←——→ Keypad functions scanning
(22) ←——→ Keypad functions scanning
(23) ←——→ Keypad functions scanning
(24) ←——→ Keypad functions scanning
(25) ←——→ Keypad functions scanning
(26) ←——→ Keypad functions scanning
(27) ←——→ Keypad functions scanning
(28) ←——→ Keypad functions scanning
(29) ←——→ Keypad functions scanning
(30) N/c
(31) ——→ Infra-red transmitting diode driver
(32) ←—— 3 V supply

Notes:

M50461-36P

Infra-red remote controller IC

Pins
(1) Ground
(2) Ground
(3) N/c
(4) ←——→ System oscillator 455 kHz
(5) ←——→ System oscillator 455 kHz
(6) N/c
(7) N/c
(8) ←——→ Keypad functions scanning
(9) N/c
(10) N/c
(11) N/c
(12) ←——→ Keypad functions scanning
(13) ←——→ Keypad functions scanning
(14) ←——→ Keypad functions scanning
(15) ←——→ Keypad functions scanning
(16) N/c
(17) N/c
(18) ←——→ Keypad functions scanning
(19) ←——→ Keypad functions scanning
(20) ←——→ Keypad functions scanning
(21) ←——→ Keypad functions scanning
(22) ←——→ Keypad functions scanning
(23) ←——→ Keypad functions scanning
(24) ←——→ Keypad functions scanning
(25) ←——→ Keypad functions scanning
(26) ←——→ Keypad functions scanning
(27) ←——→ Keypad functions scanning
(28) ←——→ Keypad functions scanning
(29) ←——→ Keypad functions scanning
(30) N/c
(31) ——→ Infra-red transmitting diode driver
(32) ←—— 3V supply

Notes:

M50461-056

M50461-056

Infra-red remote controller IC

Pins
(1) Ground
(2) Ground
(3)
(4) ←——→ System oscillator 455 kHz
(5) ←——→ System oscillator 455 kHz
(6) ←——→ Keypad functions scanning
(7) ←——→ Keypad functions scanning
(8) ←——→ Keypad functions scanning
(9) ←——→ Keypad functions scanning
(10) ←——→ Keypad functions scanning
(11) ←——→ Keypad functions scanning
(12) ←——→ Keypad functions scanning
(13) ←——→ Keypad functions scanning
(14) ←——→ Keypad functions scanning
(15) ←——→ Keypad functions scanning
(16) N/c
(17) N/c
(18) ←——→ Keypad functions scanning
(19) ←——→ Keypad functions scanning
(20) N/c
(21) ←——→ Keypad functions scanning
(22) ←——→ Keypad functions scanning
(23) ←——→ Keypad functions scanning
(24) ←——→ Keypad functions scanning
(25) ←——→ Keypad functions scanning
(26) ←——→ Keypad functions scanning
(27) ←——→ Keypad functions scanning
(28) ←——→ Keypad functions scanning
(29) ←——→ Keypad functions scanning
(30) ←——→ Keypad functions scanning
(31) ——→ Infra-red transmitting diode driver
(32) ←—— 6 V supply

1 32 16

Variation

(7), (12), (18) N/c

(Continued opposite)

30 Pin DLL version variation

(7) N/c
(18) N/c
(20) ←——→ Keypad functions scanning
(29) ——→ Infra-red transmitter diode driver
(30) ←—— 3 V supply

Notes:

M50461-113FP

Infra-red remote controller IC

Pins
(1) Ground
(2) Ground
(3)
(4) ←——→ System oscillator 455 kHz
(5) ←——→ System oscillator 455 kHz
(6) ——→ Service switch
(7) ←——→ Keypad functions scanning
(8) ←——→ Keypad functions scanning
(9) ←——→ Keypad functions scanning
(10) N/c
(11) N/c
(12) ←——→ Keypad functions scanning
(13) N/c
(14) ←——→ Keypad functions scanning
(15) ←——→ Keypad functions scanning
(16) N/c
(17) N/c
(18) ←——→ Keypad functions scanning
(19) N/c
(20) ←——→ Keypad functions scanning
(21) ←——→ Keypad functions scanning
(22) ←——→ Keypad functions scanning
(23) ←——→ Keypad functions scanning
(24) ←——→ Keypad functions scanning
(25) ←——→ Keypad functions scanning
(26) ←——→ Keypad functions scanning

(Continued overleaf)

(27) ←——→ Keypad functions scanning
(28) ←——→ Keypad functions scanning
(29) ←——→ Keypad functions scanning
(30) N/c
(31) ——→ Infra-red transmitting diode driver
(32) ←—— 3 V supply

Notes:

M50462P

Infra-red remote controller IC

Pins
(1) Ground
(2) ←——→ System oscillator 393 kHz
(3) ←——→ System oscillator 393 kHz
(4) N/c
(5) ←——→ Keypad functions scanning
(6) ←——→ Keypad functions scanning
(7) ←——→ Keypad functions scanning
(8) ←——→ Keypad functions scanning
(9) ←——→ Keypad functions scanning
(10) ←——→ Keypad functions scanning
(11) ←——→ Keypad functions scanning
(12) ←——→ Keypad functions scanning
(13) ←——→ Keypad functions scanning
(14) ←——→ Keypad functions scanning
(15) ←——→ Keypad functions scanning
(16) ←——→ Keypad functions scanning
(17) ←——→ Keypad functions scanning
(18) ←——→ Keypad functions scanning
(19) ←——→ Keypad functions scanning
(20) ←——→ Keypad functions scanning
(21) ←——→ Keypad functions scanning
(22) N/c
(23) ——→ Infra-red transmitter diode driver
(24) ←—— 3 V supply

1 24 12

Variation

(21) N/c

M50467-001P

M50467-021FP

M50467-133FP

Infra-red remote controller IC

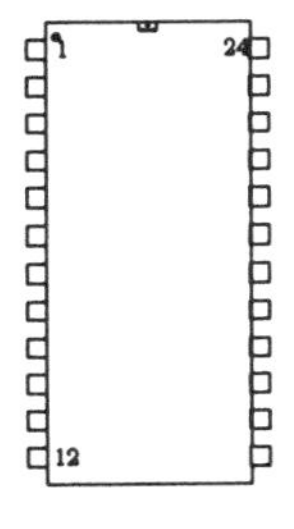

Pins
(1) Ground
(2) ←——→ System oscillator 455 kHz
(3) ←——→ System oscillator 455 kHz
(4) N/c
(5) ←——→ Keypad functions scanning
(6) ←——→ Keypad functions scanning
(7) ←——→ Keypad functions scanning
(8) ←——→ Keypad functions scanning
(9) ←——→ Keypad functions scanning
(10) ←——→ Keypad functions scanning
(11) ←——→ Keypad functions scanning
(12) ←——→ Keypad functions scanning
(13) ←——→ Keypad functions scanning
(14) ←——→ Keypad functions scanning
(15) ←——→ Keypad functions scanning
(16) ←——→ Keypad functions scanning
(17) ←——→ Keypad functions scanning
(18) ←——→ Keypad functions scanning
(19) ←——→ Keypad functions scanning
(20) ←——→ Keypad functions scanning
(21) N/c
(22) Ground
(23) ——→ Infra-red transmitting diode driver
(24) ←—— 3 V or 6 V supply

Variation

Pins
(4) ——→ TV/ VCR select
(21) ←——→ Keypad functions scanning

Notes:

M50560-002FP

M50560-002FP

Infra-red remote controller IC

Pins
(1) Ground
(2) Ground
(3)
(4) ←——→ System oscillator 455 kHz
(5) ←——→ System oscillator 455 kHz
(6)
(7) ←——→ Keypad functions scanning
(8) ←——→ Keypad functions scanning
(9) ←——→ Keypad functions scanning
(10) ←——→ Keypad functions scanning
(11) N/c
(12) N/c
(13) N/c
(14) N/c
(15) N/c
(16) N/c
(17) N/c
(18) N/c
(19) ←——→ Keypad functions scanning
(20) ←——→ Keypad functions scanning
(21) ←——→ Keypad functions scanning
(22) ←——→ Keypad functions scanning
(23) ——→ Infra-red transmitting diode driver
(24) ←—— 3 V supply

Notes:

M50747-A44SP

Control microprocessor IC

Pins

(1) ←——— 5 V supply
(2) ———→ Control signal to external digital to analogue converter
(3) ———→ Clock external digital to analogue converter
(4) ←——→ Data external digital to analogue converter
(5) ———→ Select TV = high, video = low
(6) ———→ Internal/external signal source select
(7) ———→ Sound mute
(8) ———→ Picture blanking
(9) ———→ External on screen display IC select data transfer = low
(10) ———→ External on screen display IC clock
(11) ———→ On screen display data
(12) ———→ On screen display data
(13) ———→ On screen display data
(14) ———→ On screen display data
(15) ———→ Power on/standby control
(16) ←——— Received signal ident pulse
(17) N/c
(18) ———→ VCR select = low
(19) ———→ Teletext clock
(20) N/c
(21) ←——→ Data
(22) ———→ Change audio bleep to sound circuit
(23) ←——— On screen display vertical blanking
(24) ———→ Teletext
(25) ←——— Remote control data
(26) ←——— Remote control data
(27) Chassis
(28) ←——— Power on reset = low
(29) ←——→ System oscillator 4 MHz
(30) ←——→ System oscillator 4 MHz
(31) N/c
(32) Chassis
(33) ←——— 5 V supply

(Continued overleaf)

M50747-A44SP

(34) ←———— 5 V supply
(35) Chassis
(36) Chassis
(37) Chassis
(38) ←———— Stereo sound select
(39) ————→ Bilingual sound select
(40) ————→ External TV/ VCR switch
(41) ←———→ Keypad function scanning
(42) ←———→ Keypad function scanning
(43) ←———→ Keypad function scanning
(44) ←———→ Keypad function scanning
(45) ←———→ Keypad function scanning
(46) ←———→ Keypad function scanning
(47) ←———→ Keypad function scanning
(48) ←———→ Keypad function scanning
(49) ————→ External memory mode control
(50) ————→ External memory mode control
(51) ←———→ External memory data
(52) ————→ External memory select, data transfer = low
(53) ————→ External memory mode control
(54) ————→ Clock
(55) ←———→ Data
(56) ————→ Tuner pre-scaler control
(57) N/c
(58) ————→ Tuning band select
(59) ←———— Auto channel fine tuning
(60) ————→ Stereo = high, mono = low
(61) ————→ NICAM modes select
(62) ————→ Additional sound muting
(63) ————→ Spatial sound select
(64) ————→ NICAM sound select

Variation

(58) N/c

....

Notes:

M343000 - 583SP

M343000 - 681SP

Control microprocessor IC

Pins
(1) ⟶ On screen display character
(2) ⟶ Tuning band selection UHF = low
(3) ⟶ Tuning band selection UHF = high
(4) ⟶ Change bleep tone to sound circuit
(5) ⟶ Teletext standby control
(6) ⟶ SCART socket 2 status pin 8
(7) ⟵ Power on reset = low
(8) ⟶ Auto fine channel tuning
(9) ⟵ Received signal ident sync pulse
(10) ⟶ Picture contrast control
(11) ⟶ Picture colour control
(12) ⟶ Picture brightness control
(13) ⟶ Sound volume control
(14) ⟵ Remote control data
(15) ⟶ On screen display picture blanking
(16) ⟶ Sound mute
(17) ⟶ Additional muting
(18) Chassis
(19) Chassis
(20) ⟶ Channel tuning control
(21) Chassis
(22) ⟷ Keypad function scanning
(23) ⟷ Keypad function scanning
(24) ⟷ Keypad function scanning
(25) ⟷ Keypad function scanning
(26) ⟷ Keypad function scanning & external memory IC
(27) ⟷ Keypad function scanning & external memory IC
(28) ⟷ Keypad function scanning & external memory select, low = data transfer
(29) ⟵ SCART socket 1 status pin 8
(30) ⟶ Sound I.F select
(31) ⟶ Teletext enable
(32) ⟶ Timer mode LED display = low
(33) ⟶ Power on/standby control
(34) ⟷ System oscillator
(35) ⟷ System oscillator

(Continued overleaf)

M343000 - 586SP

(36) ⟶ TV/AV select
(37) ⟶ TV/AV select
(38) ⟵ On screen display vertical sync pulse
(39) ⟵ On screen display horizontal sync pulse
(40) ⟵ On screen display oscillator
(41) ⟵ On screen display oscillator
(42) ⟵ 5 V supply

Variations

Pins
(4), (5), (15), (17) N/c
(29), (36), (37) N/c

Notes:

M343000 - 586SP

Control microprocessor IC

Pins
(1) ⟶ On screen display picture blanking
(2) ⟶ Channel tuning band selection, UHF = low, VHF-H = low, VHF-L = low
(3) ⟶ Channel tuning band selection, UHF = high, VHF-H = low, VHF-L = low
(4) ⟶ Change bleep tone to sound circuit
(5) ⟶ Teletext standby
(6) ⟵ SCART socket 1 status pin 8
(7) ⟵ Power on reset = low
(8) ⟵ Auto channel fine tuning control
(9) ⟵ Received signal ident pulse
(10) ⟶ Picture contrast control
(11) ⟶ Picture colour control
(12) ⟶ Picture brightness control
(13) ⟶ Sound volume control
(14) ⟵ Remote control data
(15) ⟶ Picture blanking
(16) ⟶ Sound mute

(Continued opposite)

(17) ——→ Additional mute
(18) Chassis
(19) Chassis
(20) ——→ Channel tuning control
(21) Chassis
(22) ←——→ Keypad function scanning
(23) ←——→ Keypad function scanning
(24) ←——→ Keypad function scanning
(25) ←——→ Keypad function scanning
(26) ←——→ Keypad function scanning
(27) ←——→ Keypad function scanning
(28) ←——→ Keypad function scanning
(29) ←—— SCART socket 2 status pin 8
(30) ——→ Sound I.F switching
(31) ——→ Teletext enable
(32) ——→ Timer mode LED display
(33) ——→ Power on/standby control
(34) ←——→ System oscillator
(35) ←——→ System oscillator
(36) N/c
(37) N/c
(38) ——→ On screen display vertical sync pulse
(39) ——→ On screen display horizontal sync pulse
(40) ←——→ On screen display oscillator
(41) ←——→ On screen display oscillator
(42) ←—— 5 V supply

Variation

(4), (5), (17), (30,) (31) N/c

Notes:

M34300N4 - 583SP

M34300N4 - 583SP

Control microprocessor IC

Pins

(1) ⟶ On screen display picture blanking
(2) ⟶ Blue on screen display
(3) ⟶ Green on screen display
(4) ⟶ Red on screen display
(5)
(6)
(7) ⟵ Power on reset = low
(8) ⟵ Channel auto fine tuning
(9) ⟵ Received signal ident pulse
(10) ⟶ Picture contrast control
(11) ⟶ Picture colour control
(12) ⟶ Picture brightness control
(13) ⟶ Sound volume control
(14) ⟵ Remote control data
(15) N/c
(16) N/c
(17) N/c
(18) N/c
(19) Chassis
(20) ⟶ TV channel tuning control
(21) Chassis
(22) ⟷ Keypad functions scanning
(23) ⟷ Keypad functions scanning
(24) ⟷ Keypad functions scanning
(25) ⟷ Keypad functions scanning
(26) ⟷ Keypad functions scanning
(27) ⟷ Keypad functions scanning
(28) ⟷ Keypad functions scanning
(29) ⟷ Keypad functions scanning
(30) ⟵ Channel auto fine tuning
(31) ⟶ Tuning band VHF/UHF select
(32) ⟶ Tuning band VHF/UHF select
(33) ⟶ Power on/standby control
(34) ⟷ System oscillator
(35) ⟷ System oscillator
(36) N/c
(37) ⟶ Sound mute on = high

(Continued opposite)

(38) ←——— On screen display vertical sync pulse
(39) ←——— On screen display horizontal sync pulse
(40) ←———→ On screen display oscillator
(41) ←———→ On screen oscillator
(42) ←——— 5 V supply

Variations

(2), (3), (4), N/c
(31), (32), N/c
(36) ———→ Additional sound/picture muting

Notes:

M34300N4 - 623SP

Control microprocessor IC

Pins
(1) ———→ On screen display picture blanking
(2) ———→ Blue on screen display
(3) ———→ Green on screen display
(4) ———→ Red on screen display
(5) ———→ Select TV = high/AV = low
(6) ←——— Vertical flyback pulse
(7) ←——— Power on reset = low
(8) ←——— Auto channel fine tuning
(9) ←——— Received signal ident pulse
(10) ———→ Sound volume control
(11) ———→ Picture colour control
(12) ———→ Picture brightness control
(13) ———→ Picture contrast control
(14) ←——— Remote control data
(15) ←——— Low = power supply rail fault detect
(16) ———→ Auto channel fine tuning on = low
(17) ———→ Power on = low/standby = high
(18) ———→ AV1/AV2 select
(19) Chassis
(20) ———→ Channel tuning control
(21) Chassis

1 42 21

(Continued overleaf)

M34300N4 - 623SP

(22)←——→ Keypad functions scanning
(23)←——→ Keypad functions scanning
(24)←——→ Keypad functions scanning
(25)←——→ Keypad functions scanning
(26)←——→ Keypad functions scanning
(27)←——→ Keypad functions scanning
(28)←——→ Keypad functions scanning
(29)←——→ Keypad functions scanning
(30)←——→ Keypad functions scanning
(31)———→ Tuning band selection
(32)———→ Tuning band selection
(33)———→ AV skew picture correction, on = low
(34)←——→ System oscillator 4 MHz
(35)←——→ System oscillator 4 MHz
(36)———→ Teletext clock control
(37)←——→ Teletext data control
(38)←——— On screen display vertical sync pulse
(39)←——— On screen display horizontal sync pulse
(40)←——→ On screen display oscillator
(41)←——→ On screen display oscillator
(42)←——— 5 V supply

Variation

(2), (4), (8) N/c
(18) Chassis
(31), (32) N/c

Notes:

M37100M8 - 583

M37100M8 - 588SP

Control microprocessor IC

Pins

(1) ←——— 5 V supply
(2) ———→ Sound volume control
(3) ———→ Picture contrast control
(4) ———→ Picture brightness control
(5) ———→ Picture colour control
(6) ———→ Channel tuning control
(7) N/c
(8) ———→ Tuning band select
(9) ———→ Tuning band select
(10) ———→ Sound mute
(11) ———→ Additional muting
(12) N/c
(13) N/c
(14) N/c
(15) ———→ Change bleep tone to sound circuit
(16) N/c
(17) ———→ SCART socket 1 status pin 8
(18) ———→ SCART socket 2 status pin 8
(19) ———→ 50 Hz/60 Hz scan system switch
(20) ←——— Auto channel fine tuning
(21) ←——— Received signal ident pulse
(22) ———→ TV /AV select, TV = high, AV1 = high, AV2 = low
(23) ———→ TV /AV select, TV = high, AV1 = low, AV2 = high
(24) ←——— Remote control data
(25) ———→ Power on /standby control
(26) Chassis
(27) ←——— Power on reset = low
(28) ←———→ System oscillator 4 MHz
(29) ←———→ System oscillator 4 MHz
(30) N/c
(31) N/c
(32) Chassis
(33) N/c
(34) ←———→ On screen display oscillator
(35) ←———→ On screen display oscillator
(36) N/c

(Continued overleaf)

(37) ⟶ System clock bus
(38) ⟷ System data bus
(39) N/c
(40) N/c
(41) ⟶ Bus control digital to analogue converter
(42) ⟷ External memory select = low transferring data
(43) ⟷ Data bus
(44) ⟶ Picture mute (blue background)
(45) ⟶ Clock keypad function scanning & external memory
(46) ⟷ Keypad function scanning
(47) ⟷ Keypad function scanning
(48) ⟷ Keypad function scanning
(49) ⟷ Keypad function scanning
(50) ⟷ Keypad function scanning
(51) ⟷ Keypad function scanning
(52) ⟷ Keypad function scanning
(53) ⟷ Keypad function scanning
(54) ⟷ Keypad function scanning
(55) ⟷ Keypad function scanning
(56) ⟷ Keypad function scanning
(57) N/c
(58) ⟵ On screen display vertical sync pulse
(59) ⟵ On screen display horizontal sync pulse
(60) ⟶ On screen display picture blanking
(61) ⟶ Red on screen display
(62) ⟶ Green on screen display
(63) ⟶ Blue on screen display
(64) N/c

Variation

(3), (4) (8), (9) N/c
(5) ⟶ NICAM select
(11), (22), (23) N/c
(36), (37), (38) N/c
(41), (61), (62), (63) N/c

Notes:

M37100M8 - 617SP

Control microprocessor IC

Pins

(1) ← 5 V supply
(2) → Picture contrast control
(3) → Picture colour control
(4) → NTSC picture hue control
(5) → Picture sharpness control
(6) → Picture brightness control
(7) → Select TV = high, AV = low
(8) → Select RGB = high
(9) ↔ Keypad function scanning
(10) ↔ Keypad function scanning
(11) ↔ Keypad function scanning
(12) ↔ Keypad function scanning
(13) ↔ Keypad function scanning
(14) N/c
(15) N/c
(16) → AV switching
(17) Chassis
(18) → SCART socket 1 status pin 8
(19) → SCART socket 2 status pin 8
(20) → Auto fine channel tuning
(21) ← Received signal ident pulse
(22) → 50 Hz/60 Hz scan system
(23) → Picture muting
(24) ← Remote control data
(25) ← On screen display vertical pulse
(26) Chassis
(27) ← Power on reset = low
(28) ↔ System oscillator
(29) ↔ System oscillator
(30) N/c
(31) N/c
(32) Chassis
(33) N/c
(34) ↔ On screen display oscillator
(35) N/c
(36) Chassis
(37) ↔ System clock bus

(Continued overleaf)

M37100M8 - 617SP

(38) ←——→ System data bus
(39) ——→ Teletext clock bus
(40) ←——→ Teletext data bus
(41) ——→ Power on/standby control
(42) ——→ NICAM decoder
(43) ——→ NICAM mono 1/mono 2 switching
(44) ——→ NICAM mono 1/mono 2 switching
(45) ——→ Stereo LED display
(46) ——→ Standby LED display
(47) ——→ NICAM LED display
(48) Chassis
(49) Chassis
(50) ——→ Auto channel frequency control
(51) ——→ Tuner sound mute
(52) ——→ On screen display picture blanking
(53) ——→ Transmission system NTSC 4.43 = high
(54) ——→ Transmission system NTSC 3.58 = high
(55) ——→ Transmission system SECAM = high
(56) ——→ Transmission system PAL 1 = low
(57) ——→ Freeze select = low
(58) ←—— On screen display vertical sync pulse
(59) ←—— On screen display horizontal sync pulse
(60) N/c
(61) ——→ Red on screen display
(62) ——→ Green on screen display
(63) ——→ Blue on screen display
(64) N/c

Variation

Pins
(53), (54), (55), (56) Chassis
(61), (63) N/c

....

Notes:

M37102M8 - A49SP

Control microprocessor IC

Pins
(1) ————→ On screen display oscillator
(2) ←———— On screen display oscillator
(3) ←———— Power failure, auto memory storage detect
(4) ←———— Channel auto frequency control
(5) ←———— Remote control data
(6) ————→ Transmission system switching
(7) ————→ Sound volume control
(8) ————→ NTSC picture hue control
(9) ————→ Picture colour control
(10) ————→ Picture brightness control
(11) ————→ Picture contrast control
(12) ————→ Picture sharpness control
(13) ————→ Sound bass control
(14) ————→ Sound treble control
(15) ————→ Picture aspect ratio control, low = 4:3, high = 16:9
(16) Chassis
(17) ————→ Multi sound effects processing 1
(18) ←———→ Keypad functions scanning
(19) ————→ Sound balance control
(20) ————→ Multi sound effects processing 2
(21) ————→ System clock bus
(22) ←———→ System data bus
(23) ————→ Teletext /RGB/TV switching
(24)
(25)
(26) ————→ External memory clock
(27) Chassis
(28)
(29) ←———— Power on reset = low
(30) ←———→ System oscillator 4 MHz
(31) ←———→ System oscillator 4 MHz
(32) Chassis
(33) ————→ Auto gain control to tuner
(34) ————→ Tuner enable
(35) ←———→ Data bus to tuner

(Continued overleaf)

M37102M8 - A49SP

(36) ——————→ Tuner clock bus
(37) ←—————— SCART socket status pin 8
(38) ←—————→ Keypad functions scanning
(39) ←—————— Received signal sync pulse
(40) ←—————→ Keypad functions scanning & external memory
(41) N/c
(42) N/c
(43) ——————→ Multi sound selection control
(44) ——————→ Multi sound selection control
(45) ——————→ Picture muting
(46) ←—————→ External memory select = low transferring data
(47) ←—————→ System diode matrix scanning
(48) ←—————→ System diode matrix scanning
(49) ——————→ Stereo LED display
(50) ——————→ Super VHS select
(51) ——————→ High = TV, low = AV
(52) ——————→ High = TV, low = AV
(53) ——————→ VTR picture skew control
(54) ——————→ Picture noise reduction circuit on = high
(55) ——————→ Timer mode on = high
(56) ——————→ Power on/standby control
(57) ——————→ On screen display picture blanking
(58) ——————→ Sound mute
(59) ——————→ Blue on screen display
(60) ——————→ Green on screen display
(61) ——————→ Red on screen display
(62) ——————→ On screen display vertical sync pulse
(63) ——————→ On screen display horizontal sync pulse
(64) ←—————— 5 V supply

Variation

(6) N/c

Notes:

M37102MB - 526SP

Control microprocessor IC

Pins
(1) ←——→ On screen display oscillator 5 MHz
(2) ←——→ On screen display oscillator 5 MHz
(3) ——→ Channel auto fine tuning switch
(4) ——→ Channel auto fine tuning control
(5) ←—— Remote control data
(6) ——→ Channel tuning control
(7) ——→ Picture colour control
(8) ——→ Sound volume control
(9) ——→ Tuning band selection, UHF = high
(10) ——→ Tuning band selection, UHF = low
(11) ——→ Transmission system selection
(12) ——→ Transmission system selection
(13) ——→ Transmission system selection
(14) ——→ Transmission system selection
(15) ——→ Transmission system selection
(16) ——→ Transmission system selection
(17) ←—— 50 Hz/60 Hz scan system
(18) ——→ VTR skew control
(19) ←—— Power supply monitoring
(20) ——→ External RGB switch
(21) ←—— Received signal sync pulse
(22) ——→ Teletext circuit control/clock
(23) ——→ digital to analogue converter circuit control signal
(24) ←——→ External memory
(25) ←——→ External memory clock
(26) ←——→ External memory data
(27) Chassis
(28) N/c
(29) ←—— Power on reset
(30) ←——→ System oscillator 4 MHz
(31) ←——→ System oscillator 4 MHz
(32) Chassis
(33) ——→ Power on/standby control
(34) ←——→ Power LED display
(35) ——→ AV1/AV2 selection
(36) ——→ TV/AV select
(37) ——→ SCART socket status pin 8

(Continued overleaf)

M37102MB - 526SP

(38) ——→ Sound, mono/stereo/bilingual switching circuits
(39) ——→ Sound, mono/stereo/bilingual switching circuits
(40) ——→ Sound, mono/stereo/bilingual switching circuits
(41) N/c
(42) N/c
(43) N/c
(44) N/c
(45) ——→ Sound mute
(46) ←——→ Keypad functions scanning
(47) ←——→ Keypad functions scanning
(48) ←——→ Keypad functions scanning
(49) ←——→ Keypad functions scanning
(50) ←——→ Keypad functions scanning
(51) ←——→ Keypad functions scanning
(52) ←——→ Keypad functions scanning
(53) ←——→ Keypad functions scanning
(54) ←——→ Keypad functions scanning
(55) ←——→ Keypad functions scanning
(56) ←——→ Keypad functions scanning
(57) ——→ On screen display picture blanking
(58) N/c
(59) ——→ Blue on screen display
(60) ——→ Green on screen display
(61) ——→ Red on screen display
(62) ←——→ On screen display vertical sync pulse
(63) ←——→ On screen display horizontal sync pulse
(64) ←—— 5 V supply

Variation

Pins
(11), (12), (13), (14), (15), (16) N/c

Notes:

M37103M4

Control microprocessor IC

Pins

(1) ←——— 5 V supply
(2) ———→ Tuning band UHF select
(3) N/c
(4) ———→ Tuning band VHF H select
(5) ———→ Tuning band VHF L select
(6) ———→ TV channel tuning
(7) ———→ Audio tone to sound output circuit
(8) ←——→ Keypad function scanning
(9) ←——→ Keypad function scanning
(10)
(11)
(12)
(13) ———→ TV = low, AV = high
(14) ———→ Timer function on/off
(15) ———→ Power on/standby control
(16)
(17) ←——→ Keypad function scanning
(18) ←——→ Keypad function scanning
(19) ←——→ Data, external memory IC
(20) ←——— Auto channel frequency control
(21) ←——— Remote control data
(22) N/c
(23) N/c
(24) ←——— Received signal ident pulse
(25) ←——— Supply rail monitor
(26) Chassis
(27) ←——— Power on reset = low
(28) ←——→ System oscillator 4 MHz
(29) ←——→ System oscillator 4 MHz
(30) ←——→ Oscillator 2, 32 kHz
(31) ←——→ Oscillator 2, 32 kHz
(32) Ground
(33) N/c
(34) ←——→ On screen display oscillator
(35) ←——→ On screen display oscillator
(36) ———→ Picture contrast control
(37) ———→ Picture Brightness control

(Continued overleaf)

(38) ——→ Picture colour control
(39) ——→ Sound volume control
(40)
(41)
(42) ——→ Picture mute
(43) ——→ Sound mute
(44) ——→ Auto channel fine tuning
(45) ——→ Audio tone to sound circuits
(46) ←——→ External memory IC data
(47) ——→ External memory IC clock
(48) ——→ External memory IC select, data transfer = low
(49) Chassis
(50)
(51)
(52) Chassis
(53)
(54) Chassis
(55) Chassis
(56) Chassis
(57) Chassis
(58) ←—— On screen display vertical sync pulse
(59) ←—— On screen display horizontal sync pulse
(60) N/c
(61) ——→ Red on screen display
(62) ——→ Green on screen display
(63) ——→ Blue on screen display
(64) ——→ On screen display picture blanking

Notes:

M37202M3

Control microprocessor IC

Pins

(1) ←——→ On screen display oscillator
(2) ←——→ On screen display oscillator
(3) ←—— 50 Hz/60 Hz scan system sensing
(4) ←—— Auto channel fine tuning control
(5) ←—— Remote control data
(6) ——→ Channel tuning control
(7) ——→ Auto channel fine tuning on/off switching
(8) ——→ Sound volume control
(9) ——→ Tuning band selection
(10) ——→ Tuning band selection
(11) ——→ Transmission system PAL/SECAM selection
(12) ——→ Transmission system PAL/SECAM selection
(13) ——→ Transmission system PAL/SECAM selection
(14) ——→ Transmission system PAL/SECAM selection
(15) ——→ Transmission system PAL/SECAM selection
(16) ←—— Received signal ident pulse
(17) ——→ 50 Hz/60 Hz scan system select
(18) ——→ VTR picture skew on/off
(19) ←—— Power supply fault detect
(20) ——→ TV/RGB switching
(21) ←——→ System clock bus
(22) ←——→ System data bus
(23) ——→ External digital to analogue converter control signal
(24) ——→ Teletext data limited clock
(25) ←——→ External digital to analogue converter circuit & memory clock/data
(26) ←——→ External digital to analogue converter & teletext data
(27) Chassis
(28) N/c
(29) ←—— Power on reset = low
(30) ——→ System clock bus
(31) ←——→ System data bus
(32) Chassis

(Continued overleaf)

(33) ——→ Power on/standby control
(34) N/c
(35) ——→ TV/AV selection
(36) ——→ TV/AV selection
(37) ←—— SCART socket status pin 8
(38) ——→ Stereo mono sound switching
(39) ——→ Stereo mono sound switching
(40) ——→ Stereo/mono sound switching
(41) ←—— NICAM transmission modes detect
(42) ←—— NICAM transmission modes detect
(43) ←—— NICAM bilingual transmission detect
(44) ←—— NICAM stereo transmission detect
(45) ——→ Sound output circuit mute
(46) ——→ NICAM circuit muting
(47) N/c
(48) ←——→ Keypad functions scanning
(49) ←——→ Keypad functions scanning
(50) ←——→ Keypad functions scanning
(51) ←——→ Keypad functions scanning
(52) ←——→ Keypad functions scanning
(53) ←——→ Keypad functions scanning
(54) ←——→ Keypad functions scanning
(55) ←——→ Keypad functions scanning
(56) ←——→ Keypad functions scanning
(57) ——→ On screen display picture blanking
(58) N/c
(59) ——→ Blue on screen display
(60) ——→ Green on screen display
(61) ——→ Red on screen display
(62) ←—— On screen display vertical sync pulse
(63) ←—— On screen display horizontal sync pulse
(64) ←—— 5 V supply

Variation

(11), (12), (14) N/c
(43), (44) Chassis

Notes:

MAB8049

Control microprocessor IC

Pins

(1) ←—— Remote control data
(2) ←——→ System oscillator 6 MHz
(3) ←——→ System oscillator 6 MHz
(4) ←—— Power on reset = low
(5) ←—— 5 V supply
(6) N/c
(7) Chassis
(8) N/c
(9) N/c
(10) N/c
(11) ——→ External channel tuning IC, 400 kHz clock reference
(12) ——→ Sound, stereo/mono decoder control
(13) ——→ Sound, speech/music select
(14) ——→ Selected sound function LED indicator
(15) ——→ Channel program 49 select
(16) ——→ AV select
(17) ——→ TV tuning band select VHF L
(18) ——→ TV tuning band select VHF H
(19) ——→ TV tuning band select UHF
(20) Chassis
(21) ——→ Auto fine channel tuning on/off
(22) ——→ Mute
(23) ——→ Power on/standby control
(24) N/c
(25) N/c
(26) ←—— 5 V supply
(27) ←——→ Keypad functions scanning
(28) ←——→ Keypad functions scanning
(29) ←——→ Keypad functions scanning
(30) ←——→ Keypad functions scanning
(31) ←——→ Keypad functions scanning
(32) ←—— External memory IC data
(33) ——→ External memory IC select, data transfer = low
(34) ←——→ Serial data bus
(35) ——→ External memory IC clock
(36) ——→ Data control
(37) ——→ External memory IC data

(Continued overleaf)

(38) Chassis
(39) ← Received signal ident pulse
(40) ← 5 V supply

Notes:

MAB8441P

Control microprocessor IC

Pins
(1) → Seven segment LED display
(2) ↔ Serial data bus
(3) → Serial clock bus
(4) ↔ Keypad functions scanning
(5) ↔ Keypad functions scanning
(6) ↔ Keypad functions scanning
(7) → Sound ambience effect, high = on
(8) → TV tuning band select
(9) → TV tuning band select
(10) → Seven segment LED display
(11) → Seven segment LED display
(12) ← Remote control data
(13) → TV tuning band select
(14) Chassis
(15) ↔ System oscillator
(16) ↔ System oscillator
(17) ← Power on reset = low
(18) ↔ Keypad functions scanning & seven segment LED display
(19) ↔ Keypad functions scanning & seven segment LED display
(20) ↔ Keypad functions scanning & seven segment LED display
(21) ↔ Keypad functions scanning & seven segment LED display
(22) ↔ Keypad functions scanning & seven segment LED display

(Continued opposite)

(23) ←——→ Keypad functions scanning & seven segment LED display
(24) ←——→ Keypad functions scanning & seven segment LED display
(25) N/c
(26) ——→ Power on / standby control
(27) ——→ Sound effects, music = low, speech = high
(28) ←—— 5V supply

Variation

(7).Chassis
(8), (9), (13) N/c

Notes:

MAB8441P

Control microprocessor IC (As used in Teletext circuit)

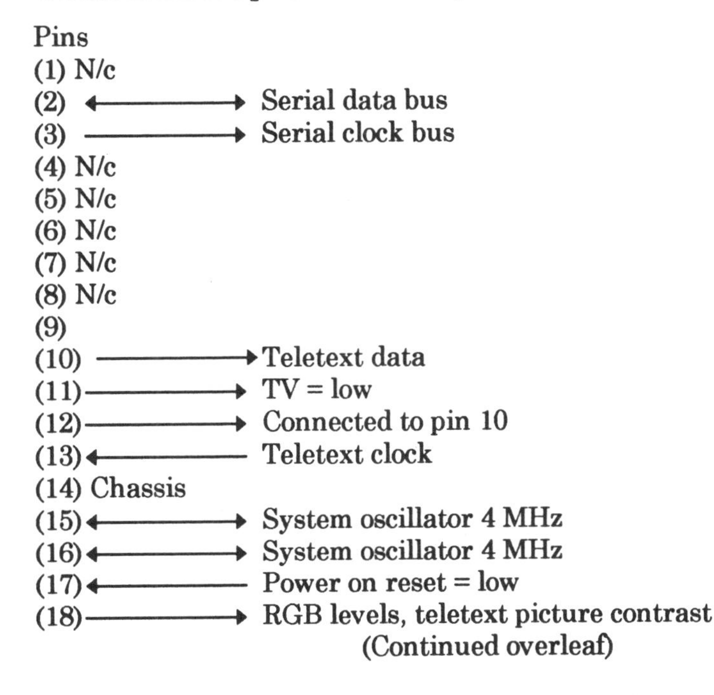

Pins
(1) N/c
(2) ←——→ Serial data bus
(3) ——→ Serial clock bus
(4) N/c
(5) N/c
(6) N/c
(7) N/c
(8) N/c
(9)
(10) ——→ Teletext data
(11) ——→ TV = low
(12) ——→ Connected to pin 10
(13) ←—— Teletext clock
(14) Chassis
(15) ←——→ System oscillator 4 MHz
(16) ←——→ System oscillator 4 MHz
(17) ←—— Power on reset = low
(18) ——→ RGB levels, teletext picture contrast

(Continued overleaf)

(19) ——→ RGB levels, teletext picture contrast
(20) ——→ RGB levels, teletext picture contrast
(21) ——→ RGB levels, teletext picture contrast
(22) Chassis
(23) Chassis
(24) Chassis
(25) Chassis
(26) N/c
(27) N/c
(28)

Notes:

MAB8441P - TO48

Control microprocessor IC

Pins
(1) ←—— 5 V supply
(2) ←——→ Serial data bus
(3) ——→ Serial clock bus
(4) ←——→ Keypad functions scanning
(5) ←——→ Keypad functions scanning
(6) ←——→ Keypad functions scanning
(7) ←——→ Keypad functions scanning
(8) ←——→ Keypad functions scanning
(9) ←——→ Keypad functions scanning
(10) ←——→ Keypad functions scanning
(11) ←——→ Keypad functions scanning
(12) ←—— Remote control data
(13)
(14) Chassis
(15) ←——→ System oscillator 4 MHz
(16) ←——→ System oscillator 4 MHz
(17) ←—— Power on reset = low
(18) ←—— Power switch pulse contact
(19) N/c
(20) N/c

(Continued opposite)

(21) N/c
(22) N/c
(23) N/c
(24) N/c
(25) ⟶ Standby & remote command LED indicator
(26) N/c

(27) N/c
(28) ⟵ 5 V supply

Variation

(7) N/c
(8) Chassis
(9) N/c
(10) ⟶ Channel LED indicator & seven segment LED display
(11) ⟶ Seven segment LED display
(18) ⟷ Keypad functions scanning & seven segment LED display
(19) ⟷ Keypad functions scanning & seven segment LED display
(24) ⟷ Keypad functions scanning & seven segment LED display
(26) ⟵ Power switch pulse contact

Notes:

MAB8461 - PW146

MAB8461 - PW146

Control microprocessor IC

Pins
(1) ————→ Data bus, external on screen display IC
(2) ←———→ Serial data bus
(3) ————→ Serial clock bus
(4) ←———→ Keypad functions scanning
(5) ←———→ Keypad functions scanning
(6) ←———→ Keypad functions scanning
(7) ←———→ Keypad functions scanning
(8) ←———→ Keypad functions scanning
(9) ————→ LED indicator driver
(10) N/c
(11) N/c
(12) ←———— Remote control data
(13) ←———— Power supply rails monitoring
(14) Chassis
(15) ←———→ System oscillator 4 MHz
(16) ←———→ System oscillator 4 MHz
(17) ←———— Power on reset = low
(18) ←———— Power switch pulse contact = low
(19) ————→ External on screen display IC select, data transfer = low
(20) N/c
(21) ————→ AV selection switching
(22) ————→ Top/bottom select, picture in picture IC
(23) ————→ Left/right select, picture in picture IC
(24) ————→ Picture in picture on = low
(25) ————→ Picture still select = low, picture in picture IC
(26) ————→ External on screen display IC select, data transfer = low
(27) ————→ External on screen display IC clock bus
(28) ←———— 5 V supply

Variation

(20), (21), (22), (23) N/c
(24), (25) Chassis

Notes:

MAB8461 - PW158

Control microprocessor IC

Pins
(1) ———→ Data, seven segment display driver
(2) ←——→ Data bus, external memory IC & circuits
(3) ———→ Clock bus, external memory IC & circuits
(4) ←——→ Keypad functions scanning
(5) ←——→ Keypad functions scanning
(6) ←——→ Keypad functions scanning
(7) ←——→ Keypad functions scanning
(8) ←——→ Keypad functions scanning
(9) ←——→ Keypad functions scanning
(10) ←——→ Keypad functions scanning
(11) ———→ Sound mute
(12) ←——— Remote control data
(13) N/c
(14) Chassis
(15) ←——→ System oscillator 4 MHz
(16) ←——→ System oscillator 4 MHz
(17) ←——— Power on reset = low
(18) ———→ Power on/standby control
(19) ———→ Transmission system, PAL = low, SECAM = high
(20) ———→ Hyperband (VHF to UHF 21-69) select
(21) ———→ Sound effect ambience, on = low
(22) ———→ External AV1 switching control
(23) ———→ External AV2 switching control
(24) Chassis
(25) ———→ Sound effect, music = high, speech = low
(26) N/c
(27)
(28) ←——— 5 V supply

Variation

Pin
(1), (11), (25) Chassis
(20) N/c

Notes:

MC68HC04

MC68HC04

Infra-red remote controller IC

Pins
(1) Ground
(2) ← 5 V supply
(3) ↔ System oscillator 4MHz
(4) ↔ System oscillator 4MHz
(5) ↔ Keypad functions scanning
(6) ↔ Keypad functions scanning
(7) ↔ Keypad functions scanning
(8) ↔ Keypad functions scanning
(9) ↔ Keypad functions scanning
(10) ↔ Keypad functions scanning
(11) ↔ Keypad functions scanning
(12) ↔ Keypad functions scanning
(13) ↔ Keypad functions scanning
(14) ↔ Keypad functions scanning
(15) ↔ Keypad functions scanning
(16) ↔ Keypad functions scanning
(17) ↔ Keypad functions scanning
(18) ↔ Keypad functions scanning
(19) ↔ Keypad functions scanning
(20) → Infra-red transmitting diode driver

Notes:

MC68HC05

Control microprocessor IC

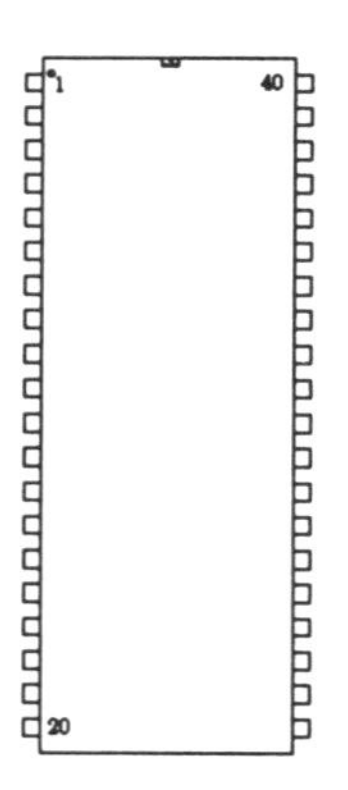

Pin
(1) ← Power on reset = low
(2) ← Remote control data
(3) N/c
(4) → VCR sync switch
(5) → TV = high, AV = low
(6) → SCART socket 1/2 select
(7) → RGB picture blanking
(8) → Picture blanking mute
(Continued opposite)

(9) ⟶ NICAM/mono 1/mono 2 select

(10) ⟶ NICAM/TV sound select

(11) ⟶ Satellite tuner enable

(12) ⟷ Keypad function scanning & 7 segment display

(13) ⟷ Keypad function scanning & 7 segment display

(14) ⟷ Keypad function scanning & 7 segment display

(15) ⟷ Keypad function scanning & 7 segment display

(16) ⟶ 7 segment display

(17) ⟶ 7 segment display

(18) ⟶ 7 segment display

(19) ⟶ 7 segment display

(20) Chassis

(21) ⟶ 7 segment display

(22) ⟶ 7 segment display

(23) ⟶ LED display

(24) ⟶ LED display

(25) ⟷ Keypad function scanning

(26) ⟶ Power on/standby control

(27) ⟶ AV sync switch

(28)

(29) ⟶ System clock bus

(30) ⟷ System data bus

(31)

(32)

(33) N/c

(34) ⟵ SCART socket status pin 8

(35) N/c

(36)

(37) ⟵ Vertical flyback pulse

(38) ⟷ System oscillator 4 MHz

(39) ⟷ System oscillator 4 MHz

(40) ⟵ 5 V supply

Notes:

MC68HC04P3

Infra-red remote controller IC

Pins
(1) Ground
(2) ←——→ Keypad functions scanning
(3) ←—— 4V5 supply
(4) ←——→ System oscillator 4.43 MHz
(5) ←——→ System oscillator 4.43 MHz
(6) Ground
(7) ——→ Infra-red transmitting diode driver
(8) ——→ To pin 13 via switch
(9) ——→ To pin 13 via switch
(10) ——→ To pin 13 via switch
(11) ——→ To pin 13 via switch
(12) N/c
(13) ←—— Switched data from pins 8, 9, 10 & 11
(14) ——→ Infra-red transmitting diode driver
(15) ←——→ Keypad functions scanning
(16) ←——→ Keypad functions scanning
(17) ←——→ Keypad functions scanning
(18) ←——→ Keypad functions scanning
(19) ←——→ Keypad functions scanning
(20) ←——→ Keypad functions scanning
(21) ←——→ Keypad functions scanning
(22) ←——→ Keypad functions scanning
(23) ←——→ Keypad functions scanning
(24) ←——→ Keypad functions scanning
(25) ←——→ Keypad functions scanning
(26) ←——→ Keypad functions scanning
(27) ←——→ Keypad functions scanning
(28) ←—— Power on reset

Variation

(6), (9) N/c
(7) ——→ Infra-red transmitting diode driver
(10) ——→ Infra-red transmitting diode driver

Notes:

MC6805

Control microprocessor IC

Pins
(1) Chassis
(2) ←——— Power on reset = low
(3) ←——— Power start up circuit feedback
(4) ←——— 5 V supply
(5) ←———→ System oscillator 4 MHz
(6) ←———→ System oscillator 4 MHz
(7) ←——— 5 V supply
(8) N/c
(9) ←——— SECAM LED display on = low
(10) ———→ SECAM & PAL LED display 5 V supply
(11) ←——— PAL LED display on = low
(12)
(13) ←——— Stereo LED display on = low
(14) ←——— Received signal ident pulse
(15) ———→ Power stages control
(16) ———→ NICAM clock bus
(17) ←———→ Teletext data bus
(18) ←———→ Teletext clock bus
(19) ———→ PAL/SECAM transmission system select
(20) ———→ PAL/SECAM transmission system select
(21) ———→ System data bus
(22) ———→ TV, I.F enable
(23) ———→ System clock bus
(24) ———→ Teletext data limited clock control
(25) ———→ Seven segment display
(26) ———→ Seven segment display
(27) ———→ Seven segment display
(28) ———→ Seven segment display
(29) ———→ Seven segment display
(30) ———→ Seven segment display
(31) ———→ Seven segment display
(32) ———→ Seven segment display
(33) ←———→ Keypad function scanning
(34) ←———→ Keypad function scanning
(35) ←———→ Keypad function scanning
(36) ←———→ Keypad function scanning
(37) ←———→ Keypad function scanning

(Continued overleaf)

(38) ←——→ Keypad function scanning
(39) ←—— SCART socket status pin 8
(40) ——→ Teletext enable

Notes:

MC144105

Infra-red remote controller IC

Pins
(1) ←——→ Keypad functions scanning
(2) ←——→ Keypad functions scanning
(3) ←——→ Keypad functions scanning
(4) ——→ Infra-red transmitting diode driver
(5) ←——→ Keypad functions scanning
(6) ←——→ Keypad functions scanning
(7) ←——→ Keypad functions scanning
(8) ←——→ Keypad functions scanning
(9) ←——→ Keypad functions scanning
(10) ←——→ Keypad functions scanning
(11) ←——→ Keypad functions scanning
(12) ←——→ Keypad functions scanning
(13) ←——→ Keypad functions scanning
(14) ←——→ Keypad functions scanning
(15) ←——→ Keypad functions scanning
(16) ←——→ Keypad functions scanning
(17) ——→ System oscillator 485 kHz
(18) ←——→ Keypad functions scanning
(19) ←——→ Keypad functions scanning
(20) ←—— 9 V supply

Variation

Pins
(1), (10) Ground

Notes:

MN6014A

Infra-red remote controller IC

Pins
(1) ←——— 3 V supply
(2)
(3) N/c
(4) ←——→ System oscillator 456 kHz
(5) ←——→ System oscillator 456 kHz
(6) ———→ Infra-red transmitting diode driver
(7) N/c
(8) ←——→ Keypad functions scanning
(9) ←——→ Keypad functions scanning
(10) N/c
(11) ←——→ Keypad functions scanning
(12) ←——→ Keypad functions scanning
(13) ←——→ Keypad functions scanning
(14) ←——→ Keypad functions scanning
(15) ←——→ Keypad functions scanning
(16) ←——→ Keypad functions scanning
(17) ←——→ Keypad functions scanning
(18) ←——→ Keypad functions scanning
(19) ←——→ Keypad functions scanning
(20) ←——→ Keypad functions scanning
(21) ←——→ Keypad functions scanning
(22) Ground

Notes:

MN6027A

Infra-red remote controller IC

Pins
(1) ←——— 3 V supply
(2) ←——→ Keypad function scanning
(3) ←——→ Keypad function scanning
(4) ←——→ Keypad function scanning

(Continued overleaf)

(5) ←——→ Keypad function scanning
(6) ←——→ Oscillator 455 kHz
(7) ←——→ Oscillator 455 kHz
(8) ——→ TV/ VCR select switch
(9) ——→ Infra red transmitting diode driver
(10) ←——→ Keypad function scanning
(11) ←——→ Keypad function scanning
(12) ←——→ Keypad function scanning
(13) ←——→ Keypad function scanning
(14) ←——→ Keypad function scanning
(15) ←——→ Keypad function scanning
(16) ←——→ Keypad function scanning
(17) ←——→ Keypad function scanning
(18) Chassis

Notes:

MN6030

Infra-red remote controller IC

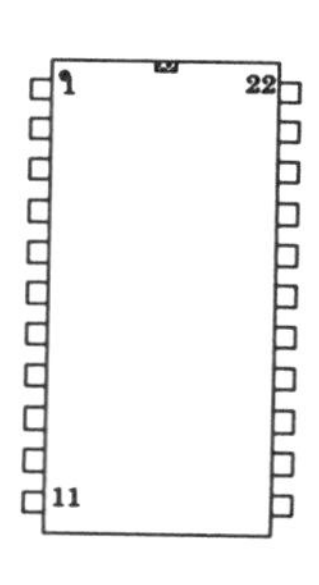

Pins
(1) ←—— 3 V supply
(2)
(3) N/c
(4) ←——→ System oscillator 455 kHz
(5) ←——→ System oscillator 455 kHz
(6) ——→ Infra-red transmitting diode driver
(7) N/c
(8) ←——→ Keypad functions scanning
(9) ←——→ Keypad functions scanning
(10) N/c
(11) ←——→ Keypad functions scanning
(12) ←——→ Keypad functions scanning
(13) ←——→ Keypad functions scanning
(14) ←——→ Keypad functions scanning
(15) ←——→ Keypad functions scanning
(16) ←——→ Keypad functions scanning
(17) ←——→ Keypad functions scanning
(18) ←——→ Keypad functions scanning

(Continued opposite)

(19) ←——→ Keypad functions scanning
(20) ←——→ Keypad functions scanning
(21) ←——→ Keypad functions scanning
(22) Ground

Notes:

MN8303

Infra-red remote controller IC

Pins
(1) Chassis
(2) Chassis
(3) ←——→ Serial data bus
(4) ——→ Standby mode LED indicator
(5) N/c
(6) ←——→ Keypad function scanning
(7) ←——→ Keypad function scanning & power switch pulse contact
(8) ←——→ Keypad function scanning
(9) ←——→ Keypad function scanning
(10) ←——→ Keypad function scanning
(11) ←——→ Keypad function scanning
(12)
(13) Chassis
(14) ←—— 5 V supply
(15) ←—— 5 V supply
(16) ←—— 5 V supply
(17) N/c
(18) ←——→ Keypad function scanning & power switch pulse contact
(19) ←——→ Keypad function scanning
(20) ←——→ Keypad function scanning
(21) ←——→ Keypad function scanning
(22) ←——→ Keypad function scanning
(23) ←——→ Keypad function scanning
(24)
(25)

(Continued overleaf)

(26) N/c
(27) ←——— Sub bus mode control
(28) ←——— Sub clock bus

Notes:

MN14831

Control microprocessor IC

Pins
(1) ←——— 5 V supply
(2) ←——— Power on reset = low
(3) ———→ External memory IC mode control
(4) ———→ External memory IC mode control
(5) ———→ External memory IC mode control
(6) ———→ Sound mute
(7) ———→ TV channel auto fine tuning on /off switch
(8) ———→ Power on = high, standby control = low
(9) N/c
(10) N/c
(11) N/c
(12) N/c
(13) ———→ On screen picture blanking
(14) ←——— On screen display horizontal sync pulse
(15) ←——— On screen display vertical sync pulse
(16) ———→ On screen display character
(17) Chassis
(18) ←——→ System oscillator
(19) ←——→ System oscillator
(20) Chassis
(21) ———→ TV channel tuning control
(22) ———→ Sound volume control
(23) ←——→ Keypad function scanning
(24) ←——→ Keypad function scanning
(25) ←——→ Keypad function scanning
(26) ←——→ Keypad function scanning
(27) ←——→ Keypad function scanning

1 40 20

(Continued opposite)

(28) ←——→ Keypad function scanning
(29) ←——→ Keypad function scanning
(30) ←——→ Keypad function scanning
(31) ←——→ External memory IC data
(32) ←—— Remote control data
(33) ——→ TV tuning band select UHF/ VHF
(34) ——→ TV tuning band select UHF/ VHF
(35) N/c
(36) N/c
(37) ←——→ Keypad function scanning
(38) ←——→ Keypad function scanning
(39) ←——→ Keypad function scanning
(40) ←——→ Keypad function scanning

Notes:

MN15142TEB

Control microprocessor IC

Pins
(1) Chassis
(2) N/c
(3) ←—— On screen display horizontal sync pulse
(4) ←—— On screen display vertical sync pulse
(5) ——→ Sound mute
(6) ——→ Power on/standby control
(7)
(8) ←—— Power on reset = low
(9) ←—— Remote control data
(10) ←——→ Keypad functions scanning
(11) ←——→ Keypad functions scanning
(12) ←——→ Keypad functions scanning
(13) ←——→ Keypad functions scanning
(14) ←——→ Keypad functions scanning
(15) ←——→ Keypad functions scanning
(16) ←——→ Keypad functions scanning
(17) ←——→ Keypad functions scanning
(18) ←——→ Keypad functions scanning

(Continued overleaf)

(19) ←——→ Keypad functions scanning
(20) ←——→ Keypad functions scanning
(21) ——→ 50 Hz/60 Hz scan system switching
(22) ——→ On screen display picture blanking
(23) ——→ Red on screen display
(24) ——→ Green on screen display
(25) ——→ TV channel tuning
(26) ——→ TV channel auto frequency control
(27) N/c
(28) ←——→ Keypad functions scanning
(29) ←——→ Power failure settings external memory IC
(30) ——→ Picture brightness control
(31) ——→ Picture colour control
(32) ——→ Sound volume control
(33) N/c
(34) N/c
(35) ——→ TV/AV source select
(36) N/c
(37) ←——→ System oscillator
(38) ←——→ System oscillator
(39) ←—— 5 V supply
(40) ←——→ On screen display oscillator
(41) ←——→ On screen display oscillator
(42) Chassis

Variation

Pins
(21), (23) N/c
(33) ——→ TV tuning band select
(34) ——→ TV tuning band select

Notes:

MN15151

Control microprocessor IC

Pins

(1) ←——— 5 V supply
(2) ←———→ Keypad function scanning
(3) ←——— Remote control data
(4) N/c
(5) ———→ LED indicator
(6) N/c
(7) ———→ Power on reset = low
(8) N/c
(9) ———→ Power on/standby control
(10) ———→ Sound mute
(11) ←——— 5 V supply
(12) N/c
(13) ———→ TV tuning band, UHF/ VHF L/ VHF H select
(14) ———→ TV tuning band, UHF/ VHF L/ VHF H select
(15) ———→ TV tuning band, UHF/ VHF L/ VHF H select
(16) ———→ Sound volume control
(17) ———→ TV channel tuning control
(18) ←——— Power switch pulse contact
(19) ←——— TV channel auto frequency control
(20) N/c
(21)
(22) ←——— Vertical sync pulse
(23) ———→ Transmission system switching
(24) ———→ VCR picture correction
(25) ←——— 5 V supply
(26) ←——— 5 V supply
(27) ←——— 5 V supply
(28) ←——— 5 V supply
(29) ———→ Transmission system PAL/SECAM select
(30) ←——— 5 V supply
(31) ———→ Transmission system PAL/SECAM select
(32) ←——— On screen display horizontal sync pulse
(33) ———→ On screen display enable
(34) ———→ Red on screen display
(35) ———→ Green on screen display
(36) ———→ On screen display vertical sync pulse
(37) ←——— TV channel auto fine tuning

(Continued overleaf)

(38) N/c
(39) ———→ Picture brightness control
(40) ———→ Picture contrast control
(41) ———→ Picture colour control
(42) ←——— 5 V supply
(43) ←——— 5 V supply
(44) ←——→ System oscillator
(45) ←——→ System oscillator
(46) Chassis
(47) ———→ Serial clock bus
(48) ←——→ Serial data bus
(49) ←——→ Keypad function scanning
(50) ←——→ Keypad function scanning
(51) ←——→ Keypad function scanning
(52) Chassis

Notes:

MN15221JMN

Control microprocessor IC

Pins
(1) N/c
(2) N/c
(3) ———→ Linked to pins 5, 6, & 9
(4) N/c
(7) N/c
(8) Chassis
(10) ———→ TV channel auto gain control
(11) ←——→ Data
(12) ———→ Linked to pin 12????
(13) ←——— TV channel auto frequency control
(14) ←——— 5 V supply
(15) ———→ TV channel tuning control
(16) ←——— Power on reset = low
(17) ←——→ System oscillator 4.5 MHz
(18) ←——→ System oscillator 4.5 MHz
(19) ←——— Data, remote control or external controller IC

(Continued opposite)

(20) → Tuner pre-scaler
(21) ← Tuner pre-scaler oscillator sample
(22) N/c
(23) → TV tuning band select
(24) → TV tuning band select
(25) N/c
(26) N/c
(27) ← Clock bus, external controller IC
(28) ← Received signal ident pulse

Notes:

MN15245

Control microprocessor IC

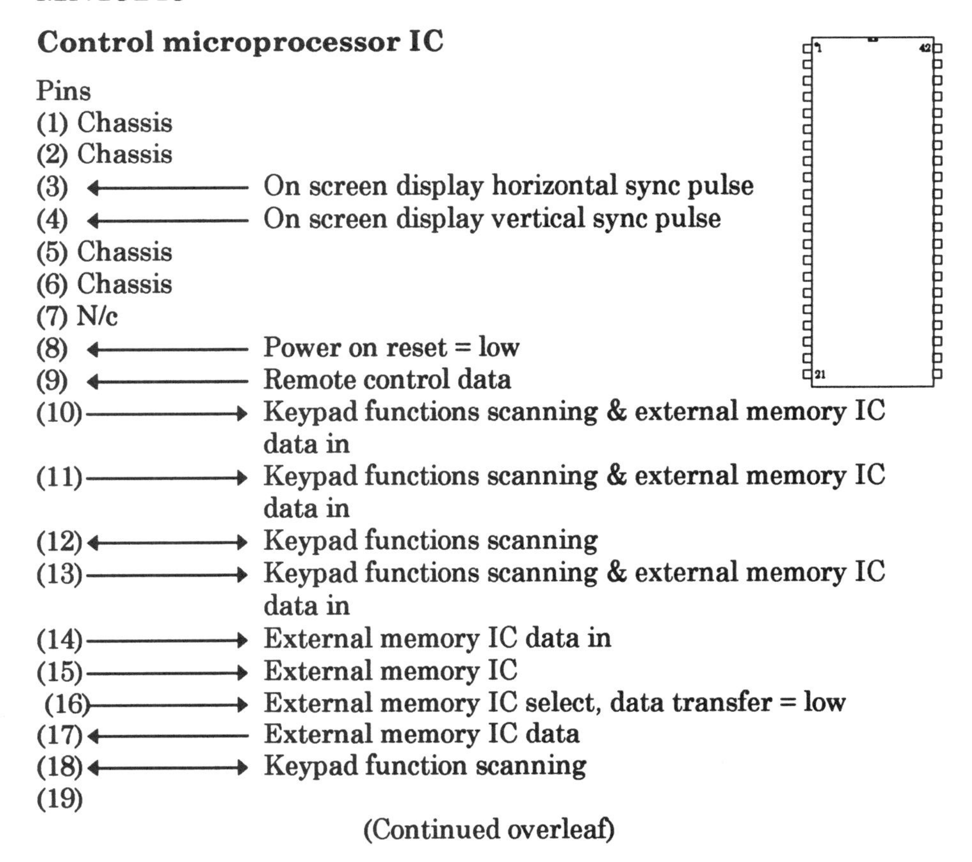

Pins
(1) Chassis
(2) Chassis
(3) ← On screen display horizontal sync pulse
(4) ← On screen display vertical sync pulse
(5) Chassis
(6) Chassis
(7) N/c
(8) ← Power on reset = low
(9) ← Remote control data
(10) → Keypad functions scanning & external memory IC data in
(11) → Keypad functions scanning & external memory IC data in
(12) ↔ Keypad functions scanning
(13) → Keypad functions scanning & external memory IC data in
(14) → External memory IC data in
(15) → External memory IC
(16) → External memory IC select, data transfer = low
(17) ← External memory IC data
(18) ↔ Keypad function scanning
(19)

(Continued overleaf)

MN15245

(20)
(21) ——→ Power on = low /standby = high
(22) ——→ On screen display character
(23) N/c
(24) N/c
(25) ——→ TV channel tuning control
(26) ——→ Tuning band select, UHF = high, VHF H = high, VHF L = low
(27) ——→ Tuning band select, UHF = high, VHF H = low, VHF L = high
(28) ——→ Tuning band select, UHF = low, VHF H = high, VHF L = high
(29) ——→ AV = high, TV = low, signal source select
(30) ——→ On screen display picture blanking
(31) ——→ Picture contrast control
(32) ——→ Picture brightness control
(33) ——→ Picture colour control
(34) ——→ Sound volume control
(35) ←——→ On screen display oscillator
(36) ←——→ Keypad function scanning
(37) ←——→ Keypad function scanning
(38) ←——→ Keypad function scanning
(39) ←—— TV channel auto fine tuning
(40) ←——→ System oscillator
(41) ←——→ System oscillator
(42) ←—— 5 V supply

Notes:

MN152811

Control microprocessor IC

Pins

(1) ← 5 V supply
(2) ← Mute, from sync separator circuit
(3) → NTSC transmission picture hue control
(4) ← Mute from signal I.F circuit
(5) → Standby LED drive
(6) → 50 Hz/60 Hz scanning system select
(7) ← Power on reset = low
(8) → SECAM/Zwietone transmission system select
(9) → TV/Teletext select
(10) → TV/AV signal source select
(11) → AV2 signal source select
(12) → Positive/negative picture modulation select (French)
(13) → Picture contrast control
(14) → Picture sharpness control
(15) → Picture colour control
(16) → Picture brightness control
(17) → RGB mode select
(18) ← TV channel auto frequency control
(19) ← Mute
(20) ↔ Keypad function scanning
(21) ↔ Keypad function scanning
(22) ← On screen display vertical sync pulse
(23) → Sound volume control
(24) → Power on/standby control
(25)
(26) ← 50 Hz/60 Hz system scanning detect
(27) → Picture mute
(28) → NTSC transmission picture hue control
(29) → Blue on screen display
(30) ← On screen display horizontal sync pulse
(31) → On screen display picture blanking
(32) → Green on screen display
(33) → Red on screen display
(34) ← Remote control data
(35) ↔ System oscillator
(36) ↔ System oscillator

(Continued overleaf)

MN15284JMP

(37) ——→ 12C clock bus, external memory & other ICs
(38) ←——→ 12C data bus, external memory & other ICs
(39) ←—— SCART socket status pin 8
(40) ——→ Super VHS select
(41) ←—— Auto contrast control
(42) Chassis

Notes:

MN15284JMP

Control microprocessor IC

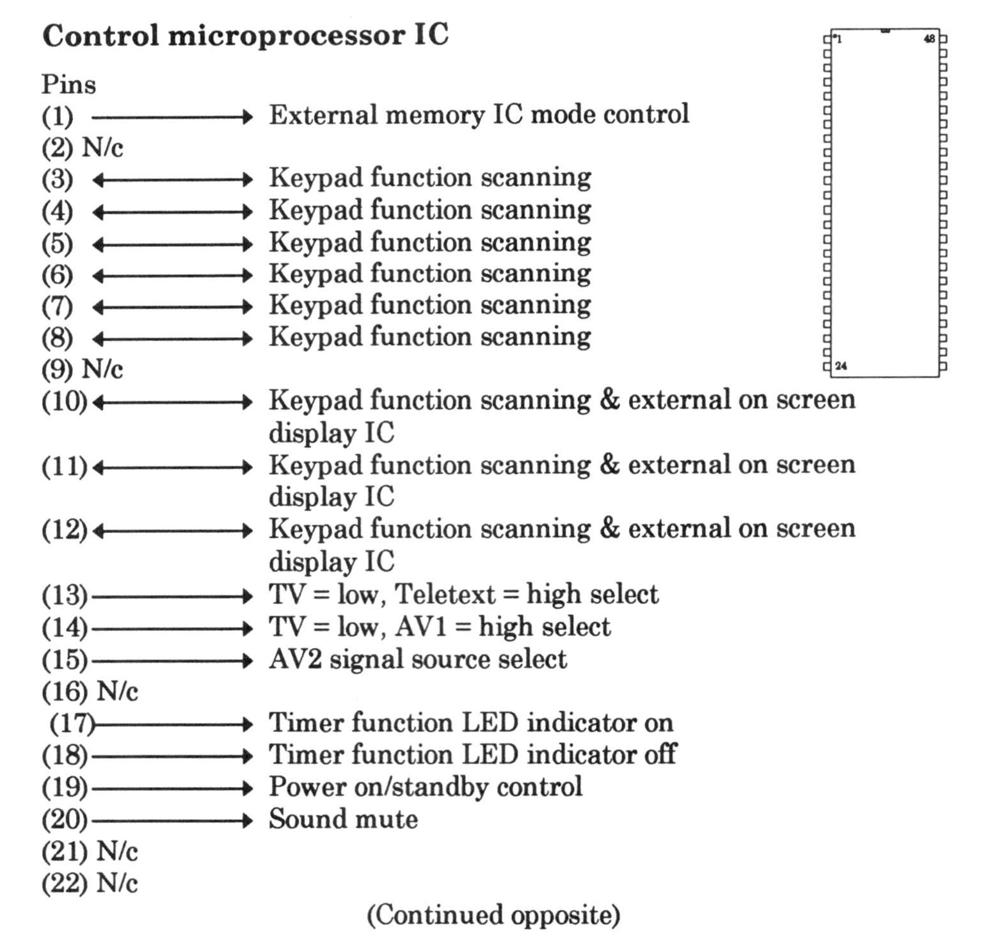

Pins
(1) ——→ External memory IC mode control
(2) N/c
(3) ←——→ Keypad function scanning
(4) ←——→ Keypad function scanning
(5) ←——→ Keypad function scanning
(6) ←——→ Keypad function scanning
(7) ←——→ Keypad function scanning
(8) ←——→ Keypad function scanning
(9) N/c
(10) ←——→ Keypad function scanning & external on screen display IC
(11) ←——→ Keypad function scanning & external on screen display IC
(12) ←——→ Keypad function scanning & external on screen display IC
(13) ——→ TV = low, Teletext = high select
(14) ——→ TV = low, AV1 = high select
(15) ——→ AV2 signal source select
(16) N/c
(17) ——→ Timer function LED indicator on
(18) ——→ Timer function LED indicator off
(19) ——→ Power on/standby control
(20) ——→ Sound mute
(21) N/c
(22) N/c

(Continued opposite)

(23) N/c
(24) ⟶ I2C serial clock bus
(25) N/c
(26) ⟷ I2C serial data bus
(27) ⟵ Remote control data
(28) ⟶ Sound volume control
(29) ⟶ Picture brightness control
(30) ⟶ Picture sharpness control
(31) ⟶ Picture contrast control
(32) N/c
(33) ⟶ Picture colour control
(34) ⟶ Sound bass control
(35) ⟶ Sound treble control
(36) ⟶ Sound balance control
(37) N/c
(38) N/c
(39) ⟵ 5 V supply
(40) N/c
(41) Chassis
(42) ⟷ System oscillator 4 MHz
(43) ⟷ System oscillator 4 MHz
(44) ⟵ Power on reset = low
(45) ⟶ Sub clock bus
(46) ⟷ Sub data bus
(47) N/c
(48) N/c

Notes:

MN15285TEH

Control microprocessor IC

Pins
(1) N/c
(2) N/c
(3) N/c
(4) ——→ Power on/standby control
(5) ——→ Sound mute
(6) ——→ TV channel auto frequency control on /off
(7) ——→ AV select
(8) N/c
(9) ——→ TV tuning band UHF, VHF H, VHF L select
(10) ——→ TV tuning band UHF, VHF H, VHF L select
(11) ——→ TV tuning band UHF, VHF H, VHF L select
(12) ←—— External memory IC data
(13) ←—— Power failure settings external memory IC
(14) ——→ Clock bus
(15) ←——→ Data bus
(16) N/c
(17) ←——→ Keypad function scanning
(18) ←——→ Keypad function scanning
(19) ←——→ Keypad function scanning
(20) ←——→ Keypad function scanning
(21) ←——→ Keypad function scanning
(22) ←——→ Keypad function scanning
(23) N/c
(24) ←——→ On screen display oscillator
(25) ←——→ On screen display oscillator
(26) ←—— 5 V supply
(27) ——→ Blue on screen display
(28) ——→ Green on screen display
(29) ——→ Red on screen display
(30) ——→ On screen display picture blanking
(31) ←—— On screen display vertical sync pulse
(32) ←—— On screen display horizontal sync pulse
(33) ←—— Remote control data
(34) ——→ Sound volume control
(35) ——→ Picture colour control
(36) ——→ Picture brightness control

(Continued opposite)

(37) ———→ Picture contrast control
(38) N/c
(39) N/c
(40) N/c
(41)
(42) ———→ TV channel tuning control
(43) ———→ External memory IC mode control
(44) ———→ TV/AV signal source select
(45) ———→ External memory IC mode control
(46) ———→ External memory IC mode control
(47) ←——→ Data bus
(48) ———→ Clock bus
(49) ←——— Power on reset = low
(50) ←——→ System oscillator
(51) ←——→ System oscillator
(52) Chassis

Variation

Pins
(7), (9), (10) N/c
(11), (14), (15) N/c
(16), (27), (47), (48) N/c

Notes:

MN1871611

MN1871611

Control microprocessor IC

Pins

(1) ←——— Remote control data
(2) ———→ NICAM sound mode select
(3) ———→ NICAM sound select
(4) ———→ Mono/stereo sound select
(5) ———→ Super VHS select
(6) ←——— SCART socket status pin 8
(7) ←——— Received signal ident pulse
(8) ———→ Transmission system, PAL/SECAM/ NTSC switching
(9) ———→ Satellite polarizer control
(10) ←——— TV auto frequency control
(11) ←——— 50 Hz /60 Hz system scan switching
(12) Chassis
(13) N/c
(14) ———→ Sound bass control
(15) ———→ Sound treble control
(16) ———→ Sound balance control
(17) ———→ Horizontal picture centre control
(18) ———→ Sound loudness control on = high
(19) ———→ Degauss circuit
(20) ———→ Sound ambience effect on = high
(21) ———→ Picture colour control
(22) ←——— 5 V supply
(23) ———→ NTSC picture hue control
(24) ———→ Picture contrast control
(25) ———→ Picture brightness control
(26) ———→ Picture sharpness control
(27) Chassis
(28) ←———→ Serial data
(29) ———→ Positive = low, negative = high, picture modulation
(30)
(31) N/c
(32) ———→ Power on /standby control
(33) N/c
(34) ———→ TV/NICAM sound select
(35) ———→ Zwietone sound
(36) N/c

1 64 32

(Continued opposite)

(37) ←——— 5 V supply
(38) ←——— 5 V supply
(39) ←——— On screen display horizontal sync pulse
(40) ———→ Sound mute 2
(41) ———→ On screen display picture blanking
(42) ———→ Blue on screen display
(43) ———→ Green on screen display
(44) ———→ Red on screen display
(45) ———→ Sound volume control
(46) ———→ Sound mute 1
(47) ———→ SCART/RGB picture blanking
(48) ———→ Satellite = high/TV = low select
(49) ———→ Multi signal source AV switching
(50) ———→ Multi signal source AV switching
(51) ———→ Transmission system PAL/SECAM/NTSC select
(52) ———→ Transmission system PAL/SECAM/NTSC select
(53) ———→ External memory IC select, data transfer = low
(54) ←——— Power on reset = low
(55) ←——— On screen display vertical sync pulse
(56) ←——— RGB blanking SCART socket pin 16
(57) ←——→ External memory IC data bus
(58) ———→ External memory IC clock bus
(59) ←——→ Serial data bus
(60) ———→ Serial clock bus
(61) ←——— 5 V supply
(62) ←——→ System oscillator
(63) ←——→ System oscillator
(64) Chassis

Notes:

MN1872419TZA

MN1872419TZA

Control microprocessor IC

Pin
(1) ←——— Remote control data
(2) ←———→ Keypad function scanning
(3) ←———→ Keypad function scanning
(4) ←———→ Keypad function scanning
(5)
(6)
(7)
(8)
(9) ←——— SCART socket status pin 8
(10) ←——— Transmission system ident
(11) ←——— TV channel auto frequency control
(12) Chassis
(13) N/c
(14) ———→ Sound bass control
(15) ———→ Sound treble control
(16) ———→ Sound balance control
(17) ———→ Sound loudness control
(18) ———→ Sound ambience effect
(19) ———→ SECAM transmission white balance
(20) ———→ SECAM L transmission
(21) ———→ SECAM L sound transmission
(22) ←——— 5 V supply
(23) ———→ Picture colour control
(24) ———→ Picture contrast control
(25) ———→ Picture brightness control
(26) ———→ Picture sharpness control
(27) Chassis
(28) ←——— NICAM data
(29) ←——— SCART socket RGB blanking status pin 16
(30) ———→ Tuner I.F mute (AV selected)
(31) ←——— Received signal ident sync pulse
(32) ←——— Super VHS/normal status
(33) ———→ high = data bus disconnected from teletext circuit, low = teletext mode (bus connect)
(34) ———→ Video mute, tri-state = high, low, high impedance
(35) ———→ PAL/SECAM auto select
(36) ———→ TV/NICAM sound select

(Continued opposite)

(37) ——→ Stereo/mono sound select
(38) ←—— 5 V supply
(39) ←—— On screen display horizontal sync pulse
(40) ——→ Power on/standby control
(41) ——→ On screen display picture blanking
(42) ——→ Blue on screen display
(43) ——→ Green on screen display
(44) ——→ Red on screen display
(45) ——→ Sound volume control
(46) ——→ TV channel tuning control
(47) ——→ TV tuning band UHF select,
(48) ——→ TV tuning band VHF H select
(49) ——→ TV tuning band VHF L select
(50) ——→ AV source switching 1
(51) ——→ AV source switching 2
(52) ——→ RGB contrast control
(53) ——→ Mute 1
(54) ←—— Power on reset = low
(55) ←—— On screen display vertical sync pulse
(56) ——→ Mute 2
(57) ←—— 50 Hz/60 Hz scan system detect
(58) ——→ TV channel auto frequency control
(59) ←——→ 12C data bus, external memory IC & teletext
(60) ——→ 12C clock bus, external memory IC & teletext
(61) ←—— 5 V supply
(62) ←——→ On screen display oscillator
(63) ←——→ On screen display oscillator
(64) Chassis

Notes:

MN1872432TMF

Control microprocessor IC

Pins
(1) ←——— Remote control data
(2) ←———→ Keypad functions scanning
(3) ←———→ Keypad functions scanning
(4) ←———→ Keypad functions scanning
(5)
(6)
(7)
(8)
(9) ←——— SCART socket status pin 8
(10)
(11) ←——— TV channel auto frequency control
(12) Chassis
(13) N/c
(14) ———→ Sound bass control
(15) ———→ Sound treble control
(16) ———→ Sound balance control
(17) ———→ Music/speech sound response switch
(18) ———→ Ambience sound effect on = high
(19) N/c
(20) N/c
(21) N/c
(22) ←——— 5 V supply
(23) ———→ Picture colour control
(24) ———→ Picture contrast control
(25) ———→ Picture brightness control
(26) ———→ Picture sharpness control
(27) Chassis
(28)
(29) ←——— SCART socket status pin 16 RGB blanking
(30) ———→ Tuner I.F mute
(31) ←——— Received signal ident sync pulse
(32) ←——— Super VHS/normal status
(33) ———→ 12C data bus disconnect control
(34) N/c
(35) N/c
(36)
(37) N/c

1 64 32

(Continued opposite)

(38) ← 5 V supply
(39) ← Horizontal sync pulse
(40) → Power on/standby control
(41) → Picture blanking
(42) → Blue on screen display
(43) → Green on screen display
(44) → Red on screen display
(45) → Sound volume control
(46) N/c
(47) N/c
(48) N/c
(49) ← Beam current protection sensing
(50) → AV2 signal source control
(51) → AV1 signal source control
(52) → RGB blanking/AV /Teletext signal source control
(53) → Mute 1 control
(54) ← Power on reset = low
(55) ← Vertical sync pulse
(56) → Mute 2
(57) ← 50 Hz/60 Hz scan system detect
(58) Chassis
(59) ↔ Serial data bus
(60) → Serial clock bus
(61) ← 5 V supply
(62) ↔ System oscillator
(63) ↔ System oscillator
(64) Chassis

Variation

Pins
(19) → SECAM transmission white balance
(20) → SECAM L transmission/PAL
(21) → SECAM L sound transmission
(28) ← NICAM data
(35) → PAL/SECAM transmission auto switching
(36) → NICAM/FM sound
(37) → Stereo/mono switching

Notes:

MPD1986

MPD1986

Infra-red remote controller IC

Pins
(1) N/c
(2) N/c
(3) N/c
(4) N/c
(5) N/c
(6) ⟷ Keypad functions scanning
(7) ⟷ Keypad functions scanning
(8) Ground
(9) ⟵ 3 V supply
(10) ⟷ System oscillator 455kHz
(11) ⟷ System oscillator 455kHz
(12) ⟶ Infra-red transmitting diode driver
(13) ⟷ Keypad functions scanning
(14) ⟷ Keypad functions scanning
(15) ⟷ Keypad functions scanning
(16) ⟷ Keypad functions scanning

Notes:

MS0467

Infra-red remote controller IC

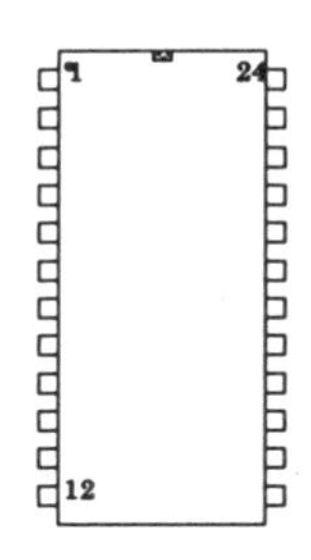

Pins
(1) Chassis
(2) ⟷ System oscillator 455 kHz
(3) ⟷ System oscillator 455 kHz
(4) N/c
(5) ⟷ Keypad function scanning
(6) ⟷ Keypad function scanning
(7) ⟷ Keypad function scanning
(8) ⟷ Keypad function scanning
(9) ⟷ Keypad function scanning
(10) ⟷ Keypad function scanning

(Continued opposite)

(11)←——→ Keypad function scanning
(12)←——→ Keypad function scanning
(13)←——→ Keypad function scanning
(14)←——→ Keypad function scanning
(15)←——→ Keypad function scanning
(16)←——→ Keypad function scanning
(17)←——→ Keypad function scanning
(18)←——→ Keypad function scanning
(19)←——→ Keypad function scanning
(20)←——→ Keypad function scanning
(21) N/c
(22) Chassis
(23)——→ Infra-red transmitting diode driver
(24)←—— 3 V or 6 V supply

Notes:

OEC1007

Control microprocessor IC

Pins
(1) ←——→ System oscillator
(2) ←——→ System oscillator
(3) ——→ TV channel tuning control
(4) Chassis
(5) ——→ External memory IC select, data transfer = low
(6) ←—— 5 V supply
(7) ——→ Sound volume
(8) ——→ TV channel auto fine tuning switch
(9) N/c
(10)——→ External memory IC clock bus
(11)
(12)←——→ Keypad function scanning & external memory IC data
(13)←——→ Keypad function scanning, external memory IC data & seven segment LED display
(14)←——→ Keypad function scanning & external memory IC data
(15)←——→ Keypad function scanning & external memory IC data

(Continued overleaf)

OEC1021A

(16)←——→ Keypad function scanning & seven segment LED display
(17)←——→ Keypad function scanning & seven segment LED display
(18)←——→ Keypad function scanning
(19)←——→ Keypad function scanning
(20)——→ Seven segment LED display
(21)——→ Seven segment LED display
(22)——→ Seven segment LED display
(23) Chassis
(24)←—— 5 V supply
(25)——→ Seven segment LED display supply
(26)——→ Seven segment LED display supply
(27)——→ Power on/standby control
(28)
(29)
(30)——→ Remote control data

Notes:

OEC1021A

Control microprocessor IC

Pins
(1)
(2)
(3) ←——→ Keypad function scanning
(4) ←——→ Keypad function scanning
(5) N/c
(6) N/c
(7) N/c
(8) ——→ Power on/standby control
(9) ←—— SCART socket status pin 8
(10)——→ Mono sound switch
(11)——→ Stereo sound switch
(12)——→ Bilingual sound switch
(13)←—— Power on reset
(14)←—— 5 V supply
(15)←—— On screen display oscillator

1 64 32

(Continued opposite)

OEC1021A

(16) ←——→ On screen display oscillator
(17) Chassis
(18) ——→ Timer mode on
(19) ——→ Bus control
(20) ——→ Bus clock
(21) ←——→ Bus data
(22) ——→ TV channel tuning control
(23) ——→ Sound volume control
(24) ——→ Picture brightness control
(25) ——→ Picture contrast control
(26) ——→ Picture colour control
(27) ←——→ System oscillator 8 MHz
(28) ←——→ System oscillator 8 MHz
(29) ——→ Red on screen display
(30) ——→ Green on screen display
(31) ——→ Blue on screen display
(32) ——→ On screen display blanking
(33) ——→ On screen display vertical sync pulse
(34) ——→ On screen display horizontal sync pulse
(35) ←—— Received signal ident pulse
(36) ——→ External memory IC mode control
(37) ——→ External memory IC select, data transfer = low
(38) ←——→ External memory IC data
(39) ←——→ External memory IC data
(40) ——→ External memory IC clock
(41) ——→ Serial clock bus
(42) ←——→ Serial data bus
(43) ——→ TV channel tuning band UHF, VHF select
(44) ——→ TV channel tuning band UHF, VHF select
(45) ——→ TV channel tuning band UHF, VHF select
(46) ——→ TV channel tuning band UHF, VHF select
(47) N/c
(48) N/c
(49) N/c
(50) ——→ Sound mono = high, stereo = low select
(51) ——→ Transmission system PAL = low, SECAM = high
(52) ——→ Signal source, TV = high, AV = low
(53) N/c
(54) ——→ TV channel auto fine tuning switch
(55) ←——→ Keypad function scanning
(56) ←——→ Keypad function scanning
(57) ←——→ Keypad function scanning

(Continued overleaf)

OEC3008

(58) ←——→ Keypad function scanning
(59) ←——→ Keypad function scanning
(60) ←——→ Keypad function scanning
(61) ←——→ Keypad function scanning
(62) ←——→ Keypad function scanning
(63) ←—— TV channel auto fine tuning control
(64) ←—— Remote control data

Variation

(9), (10), (11), (12) Chassis
(18), (19), (20), (50), (51,) (52) N/c

Notes:

OEC3008

Control microprocessor IC

Pins
(1) N/c
(2) N/c
(3) N/c
(4) ——→ Sound mute
(5) ←——→ System oscillator 4 MHz
(6) ←——→ System oscillator 4 MHz
(7) ←—— 5 V supply
(8) ←—— Power on reset = low
(9) N/c
(10) N/c
(11) N/c
(12) N/c
(13) Chassis
(14) Chassis
(15) N/c
(16) N/c
(17) ——→ TV channel tuning up
(18) ——→ TV channel tuning down
(19) ——→ Power on/standby control

(Continued opposite)

(20) ←——— 5 V supply
(21) ———→ Power switch pulse contact
(22) ←——→ Keypad function scanning
(23) N/c
(24) N/c
(25) ←——→ Keypad function scanning
(26) ←——→ Keypad function scanning
(27) ←——— Power switch pulse contact
(28) ←——— Remote control data

Notes:

OEC6008

Control microprocessor IC

Pins
(1) ———→ TV channel tuning control
(2) ———→ Picture contrast control
(3) ———→ Picture brightness control
(4) ———→ Picture colour control
(5) ———→ Sound volume control
(6) ———→ TV tuning band UHF/ VHF select
(7) ———→ TV tuning band UHF/ VHF select
(8) ———→ Channel auto fine tuning switch
(9) ←——— Channel auto fine tuning control
(10) ———→ Clock, external memory IC & keypad function scanning
(11) ←——→ Data external memory IC & keypad function scanning
(12) ←——→ Data external memory IC & keypad function scanning
(13) ←——→ Keypad function scanning
(14) ←——→ Keypad function scanning
(15) ←——→ Keypad function scanning
(16) ←——→ Keypad function scanning
(17) ←——→ Keypad function scanning
(18) ←——→ Keypad function scanning
(19) ———→ External memory IC select, data transfer = low
(20) ———→ Timer LED indicator
(21) Chassis

(Continued overleaf)

OEC6008

(22)——————→ Power on /standby control
(23)——————→ Green on screen display
(24)——————→ Red on screen display
(25)——————→ On screen display picture blanking
(26)——————→ On screen display horizontal sync pulse
(27)——————→ On screen display vertical sync pulse
(28)←—————→ On screen display oscillator
(29)←—————→ On screen display oscillator
(30) Chassis
(31)←—————→ System oscillator 4 MHz
(32)←—————→ System oscillator 4 MHz
(33)←—————— Power on reset = low
(34)←—————— 5 V supply
(35)←—————— Remote control data
(36)←—————— Received signal ident pulse
(37)——————→ TV/AV1 signal source select
(38)——————→ TV/AV2 signal source select
(39) Chassis
(40) Chassis
(41) Chassis

Variation

Pins
(6), (7) N/c
(37), (38) Chassis

....

Notes:

OEC6020A

Control microprocessor IC

Pins
(1) ←——→ Keypad function scanning
(2) ←——→ Keypad function scanning
(3) ←——→ Keypad function scanning
(4) ←——→ Keypad function scanning
(5) ←——→ Keypad function scanning
(6) ←——→ Keypad function scanning
(7) N/c
(8) ——→ TV channel auto gain control
(9) ←—— 50 Hz/60 Hz system scanning
(10) ←—— SCART socket status pin 8
(11) N/c
(12) ——→ TV channel tuning control
(13) ——→ Sound volume control
(14) ——→ Picture brightness control
(15) ——→ Picture contrast control
(16) ——→ Picture colour control
(17) ——→ Picture sharpness control
(18) N/c
(19) ——→ Sound bass control
(20) ——→ Sound treble control
(21) ——→ Sound balance control
(22)
(23)
(24) ←—— Beam current sensing
(25)
(26) ←—— TV channel auto fine tuning
(27) ←——→ Keypad function scanning
(28) ——→ TV channels tuning band, UHF/ VHF L/ VHF H select
(29) ——→ TV channels tuning band, UHF/ VHF L/ VHF H select
(30) ——→ TV channels tuning band, UHF/ VHF L/ VHF H select
(31)
(32) Chassis
(33) ——→ Red on screen display
(34) ——→ Green on screen display
(35) ——→ Blue on screen display
(36) ——→ On screen display picture blanking
(37) ←—— On screen display horizontal sync pulse

(Continued overleaf)

OEC6020A

(38) ←——— On screen display vertical sync pulse
(39) ———→ NTSC 4.43 MHz transmission select
(40) ———→ NTSC 3.58 MHz transmission select
(41) ———→ PAL = low, SECAM = high transmission select
(42) ←——→ On screen display oscillator
(43) ←——→ On screen display oscillator
(44) Chassis
(45) ←——→ System oscillator 8 MHz
(46) ←——→ System oscillator 8 MHz
(47) ←——— Power on reset = low
(48) ←——— Supply voltage monitoring
(49) N/c
(50) N/c
(51) ←——— Remote control data
(52) ←——— Received signal ident pulse
(53) ———→ Power on/standby control
(54) ———→ TV = high, AV = low select
(55) ———→ Serial clock bus
(56) ←——→ Serial data bus
(57) ———→ Additional muting
(58) ———→ Timer off/on modes
(59) ———→ Timer off/on modes
(60) ←——→ Keypad function scanning
(61) ←——→ Keypad function scanning
(62) ←——→ Keypad function scanning
(63) ←——→ Keypad function scanning
(64) ←——→ Keypad function scanning

Variation

(9), (28), (29), (30), (39), (40), (41) Chassis
(18) ———→ NTSC picture hue control
(57), 58), (59) N/c

Notes:

OEC8024

Control microprocessor IC

Pins
(1) Chassis
(2) ←——→ System oscillator 4 MHz
(3) ←——→ System oscillator 4 MHz
(4) ←—— Power on reset = low
(5) N/c
(6) Chassis
(7) ——→ External memory IC mode control & data
(8) ——→ External memory IC mode control & data
(9) ——→ External memory IC mode control & data
(10) ——→ External memory IC mode control & data
(11) ←—— Received signal ident pulse
(12) ←—— SCART socket status pin 8
(13) Chassis
(14) ←——→ Keypad function scanning
(15) ←——→ Keypad function scanning
(16) ←——→ Keypad function scanning
(17) ←——→ Keypad function scanning
(18) ←——→ Keypad function scanning
(19) ——→ Seven segment LED display
(20) ——→ Seven segment LED display
(21) ——→ Seven segment LED display
(22) ——→ Seven segment LED display
(23) ——→ Seven segment LED display
(24) ——→ Seven segment LED display
(25) ——→ Seven segment LED display
(26) ←——→ Keypad function scanning
(27) ——→ LED indicators
(28) ——→ Seven segment LED display supply
(29) ——→ Seven segment LED display supply
(30) ——→ Power on/standby control
(31) ——→ TV = high, AV = low select
(32) ——→ PAL/SECAM transmission switching
(33) N/c
(34) ——→ TV channel auto fine tuning
(35) ——→ TV tuning band UHF/ VHF select
(36) ——→ TV tuning band UHF/ VHF select
(37) ——→ External memory IC clock

(Continued overleaf)

OEC8024

(38)———→ External memory select, data transfer = low
(39)———→ Clock bus
(40)←——→ Data bus
(41)←——— Remote control data
(42)———→ Sound volume control
(43)———→ Picture brightness control
(44)———→ Picture contrast control
(45)———→ Picture colour control
(46)———→ Sound bass control
(47)———→ Sound treble control
(48)———→ Sound balance control
(49) N/c
(50)———→ TV channel tuning control
(51) N/c
(52)←——— 5 V supply

Variation

(32), (34) N/c

Notes:

OEC8034

Control microprocessor IC

Pins

(1) Chassis
(2) ←→ System oscillator 4 MHz
(3) ←→ System oscillator 4 MHz
(4) ← Power on reset = low
(5) ←→ Serial clock bus
(6) ←→ Serial data bus
(7) → External memory IC mode control/data
(8) → External memory IC mode control/data
(9) → External memory IC mode control/data
(10) → External memory IC mode control/data
(11) → Linked to pin 41
(12) ← SCART socket status pin 8
(13) → Power switch pulse contact
(14) ←→ Keypad function scanning
(15) ←→ Keypad function scanning
(16) ←→ Keypad function scanning
(17) ←→ Keypad function scanning
(18) ←→ Keypad function scanning
(19) ←→ Keypad function scanning
(20) ←→ Keypad function scanning & seven segment LED display driver
(21) ←→ Keypad function scanning & seven segment LED display driver
(22) → Seven segment LED display driver
(23) → Seven segment LED display driver
(24) → Seven segment LED display driver
(25) → Seven segment LED display driver
(26) ←→ Keypad function scanning
(27) → LED indicators
(28) → Seven segment LED display supply
(29) → Seven segment LED display supply
(30) → Power on/standby control
(31) → TV/AV signal source select
(32) N/c
(33) → Sound effects circuit switching
(34) → TV channel auto fine tuning
(35) N/c

(Continued overleaf)

OEC8034

(36) N/c
(37) ——→ External memory IC clock
(38) ——→ External memory IC select, data transfer = low
(39) ——→ Clock bus
(40) ←——→ Data bus
(41) ←—— Remote control data
(42) ——→ Sound volume control
(43) ——→ Picture brightness control
(44) ——→ Picture contrast control
(45) ——→ Picture colour control
(46) ——→ Sound bass control
(47) ——→ Sound treble control
(48) ——→ Sound balance control
(49) N/c
(50) ——→ TV channel tuning control
(51) N/c
(52) ←—— 5 V supply

Variation

(5), (6), (31), (33), (46), (47), (48) N/c

Notes:

P83CL168

Control microprocessor IC

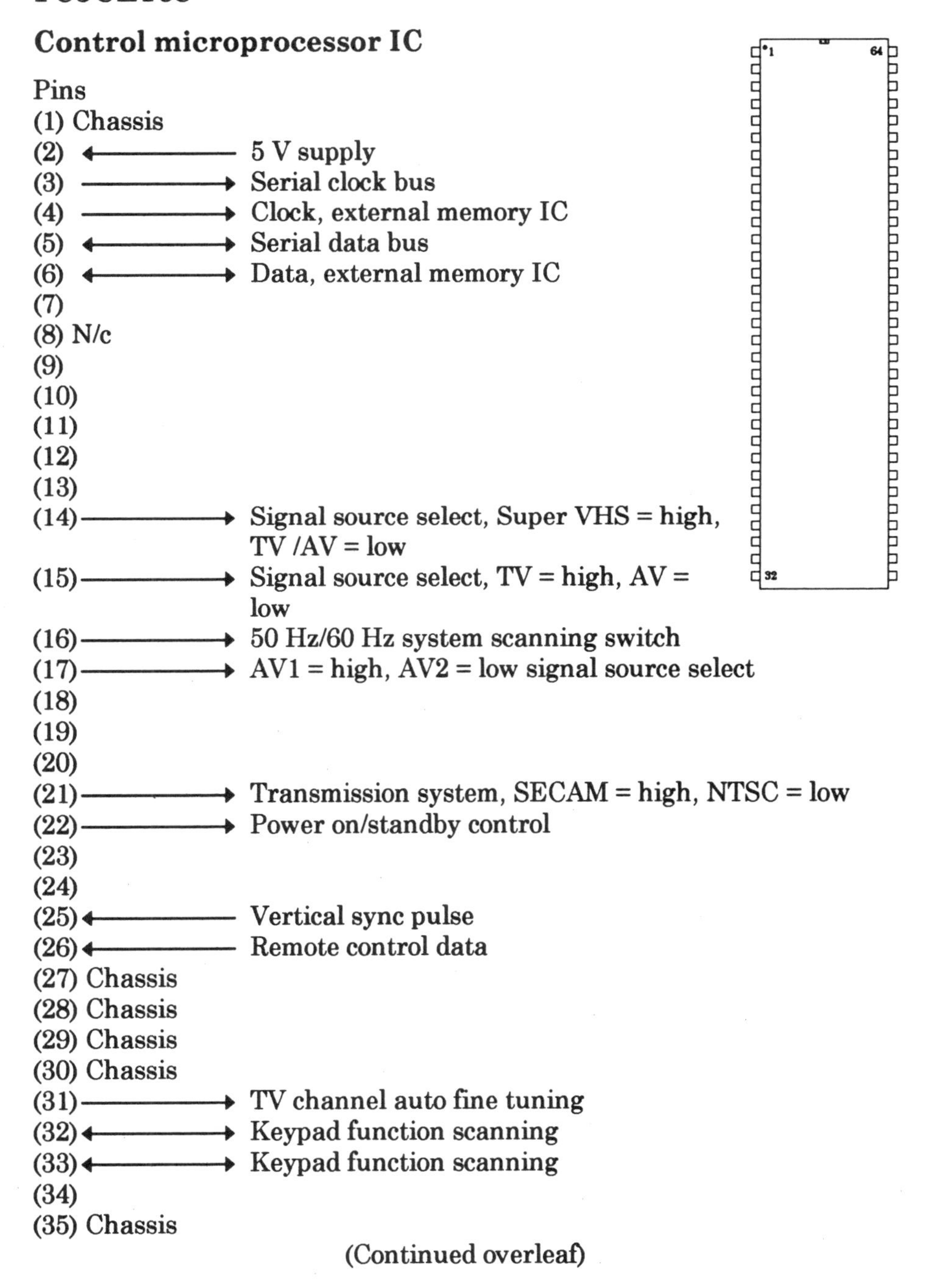

Pins
(1) Chassis
(2) ←——— 5 V supply
(3) ———→ Serial clock bus
(4) ———→ Clock, external memory IC
(5) ←——→ Serial data bus
(6) ←——→ Data, external memory IC
(7)
(8) N/c
(9)
(10)
(11)
(12)
(13)
(14) ———→ Signal source select, Super VHS = high, TV /AV = low
(15) ———→ Signal source select, TV = high, AV = low
(16) ———→ 50 Hz/60 Hz system scanning switch
(17) ———→ AV1 = high, AV2 = low signal source select
(18)
(19)
(20)
(21) ———→ Transmission system, SECAM = high, NTSC = low
(22) ———→ Power on/standby control
(23)
(24)
(25) ←——— Vertical sync pulse
(26) ←——— Remote control data
(27) Chassis
(28) Chassis
(29) Chassis
(30) Chassis
(31) ———→ TV channel auto fine tuning
(32) ←——→ Keypad function scanning
(33) ←——→ Keypad function scanning
(34)
(35) Chassis

(Continued overleaf)

P83CL168

(36) ← 5 V supply
(37) → TV /teletext picture control switching
(38) Chassis
(39) → Sound volume control
(40) → NTSC transmission picture hue control
(41) → Picture sharpness control
(42) → Picture brightness control
(43) → Picture contrast control
(44) → Picture colour control
(45) ← 5 V supply
(46) Chassis
(47) ← SCART socket AV1 status pin 8
(48)
(49) ← SCART socket AV2 status pin 8
(50) ← Received signal ident sync pulse
(51) Chassis
(52) → Sound mute
(53)
(54) ← Power on reset = low
(55) ← On screen display vertical sync pulse
(56) ← On screen display horizontal sync pulse
(57) Chassis
(58) → Blue on screen display
(59) → Green on screen display
(60) → Red on screen display
(61) → On screen display picture blanking
(62) ↔ On screen display oscillator
(63) ↔ System oscillator 12 MHz
(64) ↔ System oscillator 12 MHz

Variation

(39), (40) N/c

Notes:

PCA84C640P - 030

Control microprocessor IC

Pins
(1) ⟶ TV channel tuning control
(2) ⟶ Sound volume control
(3) ⟶ Picture brightness control
(4) ⟶ Picture colour control
(5) ⟶ Picture contrast control
(6) ⟶ Sound balance control
(7) ⟶ TV tuning band select, VHF L
(8) ⟶ TV tuning band select, VHF H
(9) ⟵ TV auto frequency control
(10) ⟶ TV tuning band select, UHF
(11) ⟶ VTR signal source select
(12) ⟶ AV signal source select
(13) ⟷ Keypad function scanning
(14) ⟷ Keypad function scanning
(15) ⟷ Keypad function scanning
(16) ⟷ Keypad function scanning
(17) ⟷ Keypad function scanning
(18) ⟷ Keypad function scanning
(19) ⟷ Keypad function scanning
(20) ⟷ Keypad function scanning
(21) Chassis
(22) ⟶ Red on screen display
(23) ⟶ Green on screen display
(24) ⟶ Blue on screen display
(25) ⟶ On screen display picture blanking
(26) ⟵ On screen display horizontal sync pulse
(27) ⟵ On screen display vertical sync pulse
(28) ⟷ On screen display oscillator
(29) ⟵ Received signal ident sync pulse
(30) Chassis
(31) ⟷ System oscillator 10 MHz
(32) ⟷ System oscillator 10 MHz
(33) ⟵ Power on reset = low
(34) ⟶ Dual language sound select
(35) ⟵ Remote control data
(36) ⟶ Stereo/mono sound mode select
(37) ⟶ Sound effects control

(Continued overleaf)

(38) ⟶ PAL/SECAM transmission select
(39) ⟶ Serial clock bus
(40) ⟷ Serial data bus
(41) ⟶ Power on/standby control
(42) ⟵ 5 V supply

Variations

Pins
(6) N/c or NTSC picture hue
(7) N/c
(10), (11), (34), (38) Chassis

Notes:

PCA84C840 - 062

Control microprocessor IC

Pins
(1) ⟶ TV channel tuning control
(2) ⟶ Sound volume control
(3) ⟶ Picture brightness control
(4) ⟶ Picture colour control
(5) ⟶ Picture contrast control
(6) ⟶ Picture sharpness control
(7) ⟶ TV tuning band UHF/ VHF select
(8) ⟶ TV tuning band UHF/ VHF select
(9) ⟵ TV channel auto fine tuning
(10) ⟶ Sound mute 1
(11) ⟵ SCART socket status pin 8
(12) ⟶ No signal sound mute
(13) ⟷ Keypad function scanning
(14) ⟷ Keypad function scanning
(15) ⟷ Keypad function scanning
(16) ⟷ Keypad function scanning
(17) ⟷ Keypad function scanning
(18) ⟷ Keypad function scanning
(19) ⟶ External RGB signal source blanking

(Continued opposite)

(20) Chassis
(21) ← Power supply rail monitoring
(22) → External RGB signal mute
(23) → Super VHS signal source select
(24) → Green on screen display
(25) → On screen display picture blanking
(26) ← On screen display horizontal sync pulse
(27) ← On screen display vertical sync pulse
(28) ↔ On screen display oscillator
(29) ← Received signal ident sync pulse
(30) Chassis
(31) ↔ System oscillator
(32) ↔ System oscillator
(33) ← Power on reset = low
(34) → External AV signal source switching
(35) ← Remote control data
(36) → TV/AV selection switching
(37) Chassis
(38) Chassis
(39) → Serial clock bus
(40) ↔ Serial data bus
(41) → Power on/standby control
(42) ← 5 V supply

Variation

(7), (8) Chassis

Notes:

PCA84C841P -506

PCA84C841P -506

Control microprocessor IC

Pins

(1) ——→ TV channel tuning control
(2) N/c
(3) ——→ Picture brightness control
(4) ——→ Picture colour control
(5) ——→ Picture contrast control
(6) ——→ Picture tint control
(7) ——→ TV tuning band UHF/ VHF select
(8) ——→ TV tuning band UHF/ VHF select
(9) ←—— TV channel auto frequency control
(10) ←—— SCART socket status pin 8
(11) ——→ VTR signal source select
(12) ——→ AV signal source select
(13) ←——→ Keypad function scanning
(14) ←——→ Keypad function scanning
(15) ←——→ Keypad function scanning
(16) ←——→ Keypad function scanning
(17) ←——→ Keypad function scanning
(18) ←——→ Keypad function scanning
(19) ←——→ Keypad function scanning
(20) ←——→ Keypad function scanning
(21) Chassis
(22) ——→ Red on screen display
(23) ——→ Green on screen display
(24) ——→ Blue on screen display
(25) ——→ On screen display picture blanking
(26) ←—— On screen display horizontal sync pulse
(27) ←—— On screen display vertical sync pulse
(28) ←——→ On screen display oscillator
(29) ←——→ On screen display oscillator
(30) Chassis
(31) ←——→ System oscillator 10 MHz
(32) ←——→ System oscillator 10 MHz
(33) ←—— Power on reset = low
(34) ←—— Received signal ident sync pulse
(35) ←—— Remote control data
(36) ——→ RGB blanking
(37) ——→ Super VHS signal select

(Continued opposite)

(38) ←——→ Keypad function scanning
(39) ——→ Serial clock bus
(40) ←——→ Serial data bus
(41) ——→ Power on/standby control
(42) ←—— 5 V supply

Notes:

PCACTV322S

Control microprocessor IC

Pins
(1) ——→ TV channel tuning control
(2) ——→ Sound volume control
(3) ——→ Picture brightness control
(4) ——→ Picture colour control
(5) ——→ Picture contrast control
(6) ——→ Sound balance control
(7) ——→ TV tuning band UHF/ VHF select
(8) ——→ TV tuning band UHF/ VHF select
(9) ——→ TV channel auto frequency control
(10) Chassis
(11) ——→ VTR signal source select
(12) ←—— SCART socket status pin 8
(13) ←——→ Keypad function scanning
(14) ←——→ Keypad function scanning
(15) ←——→ Keypad function scanning
(16) ←——→ Keypad function scanning
(17) ←——→ Keypad function scanning
(18) ←——→ Keypad function scanning
(19) ←——→ Keypad function scanning
(20) ←——→ Keypad function scanning
(21) Chassis
(22) ——→ Red on screen display
(23) ——→ Green on screen display
(24) ——→ Blue on screen display
(25) ——→ On screen display picture blanking
(26) ←—— On screen display horizontal sync pulse

(Continued overleaf)

PCF84C640

(27) ←———— On screen display vertical sync pulse
(28) ←————→ On screen display oscillator
(29) ←————→ On screen display oscillator
(30) Chassis
(31) ←————→ System oscillator 10 MHz
(32) ←————→ System oscillator 10 MHz
(33) ←———— Power on reset = low
(34) ←———— Received signal sync pulse
(35) ←———— Remote control data
(36) ————→ Sound circuits enable
(37) ————→ Sound effects circuit enable
(38)
(39) ————→ Serial clock bus
(40) ←————→ Serial data bus
(41) ————→ Power on/standby control
(42) ←———— 5 V supply

Variation

(6) N/c
(11) Chassis

Notes:

PCF84C640

Control microprocessor IC

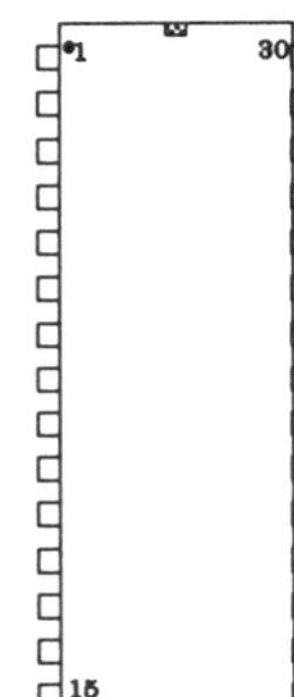

Pins
(1) ————→ Teletext circuit enable = low
(2)
(3)
(4) ————→ NICAM circuit enable = low
(5)
(6) N/c
(7) ————→ PAL transmission system select
(8) ————→ SECAM transmission system select
(9) Chassis
(10) ←———— SCART socket status pin 8

(Continued opposite)

(11)
(12) ← Beam current sensing
(13) ↔ Keypad function scanning
(14) ↔ Keypad function scanning
(15) ↔ Keypad function scanning
(16) ↔ Keypad function scanning
(17) ↔ Keypad function scanning
(18) ↔ Keypad function scanning
(19) ↔ Keypad function scanning
(20) ↔ Keypad function scanning
(21) Chassis
(22) → Red on screen display
(23) → Green on screen display
(24) → Blue on screen display
(25) → On screen display picture blanking
(26) ← On screen display horizontal sync pulse
(27) ← On screen displayed vertical sync pulse
(28) ↔ On screen display oscillator
(29) → Sound mute
(30) Chassis
(31) ↔ System oscillator 10 MHz
(32) ↔ System oscillator 10 MHz
(33) ← Power on reset = low
(34)
(35) ← Remote control data
(36) → Power on/standby control
(37) → Mono sound LED indicator
(38) → Spatial sound LED indicator
(39) → System clock bus
(40) ↔ System data bus
(41) ← Power switch pulse contact
(42) ← 5 V supply

Notes:

PCF84C640 - 021

Control microprocessor IC

Pins

(1) ——→ Power switch pulse contact
(2) ——→ Sound volume control
(3) ——→ Picture colour control
(4) ——→ Picture contrast control
(5) ——→ Picture brightness control
(6) ——→ TV channel auto fine tuning
(7) N/c
(8) N/c
(9) ——→ TV channel auto frequency control
(10) ——→ Power on/standby control
(11) ——→ Seven segment LED display
(12) ——→ Seven segment LED display
(13) ←——→ Keypad function scanning & seven segment LED display
(14) ←——→ Keypad function scanning & seven segment LED display
(15) ←——→ Keypad function scanning & seven segment LED display
(16) ←——→ Keypad function scanning & seven segment LED display
(17) ←——→ Keypad function scanning & seven segment LED display
(18) ←——→ Keypad function scanning & seven segment LED display
(19) ←——→ Keypad function scanning & seven segment LED display
(20) ←——→ Keypad function scanning & seven segment LED display
(21) Chassis
(22) ——→ Blue on screen display
(23) ——→ Green on screen display
(24) ——→ Red on screen display
(25) ——→ On screen display picture blanking
(26) ←—— On screen display horizontal sync pulse
(27) ←—— On screen display vertical sync pulse
(28) ——→ On screen display oscillator
(29) ←—— Received signal ident sync pulse

(Continued opposite)

(30) Chassis
(31)←——→ System oscillator 8 MHz
(32)←——→ System oscillator 8 MHz
(33)←—— Power on reset = low
(34)
(35)←—— Remote control data
(36)
(37)——→ AV signal source LED indicator on = low
(38)——→ RGB signal source LED indicator on = high
(39)——→ System clock IM bus
(40)←——→ System data IM bus
(41)——→ Control signal IM bus
(42)←—— 5 V supply

Variation

Note: In this example channel tuning is obtained using the clock/data bus
(7) ——→ Satellite polarizer select
(8) ——→ Satellite enable

Notes:

PLA84C641

Control microprocessor IC

Pins
(1) ——→ TV channel tuning control
(2) ——→ Sound volume control
(3) ——→ Picture brightness control
(4) ——→ Picture colour control
(5) ——→ Picture contrast control
(6) ——→ NTSC transmission picture hue control
(7) ——→ TV channels tuning band selection
(8) ——→ TV channels tuning band selection
(9) ←—— Auto fine tuning control
(10)——→ AV signal source select
(11)——→ RGB signal source switching

(Continued overleaf)

PLA84C641

(12) ←——— SCART socket status pin 8
(13) ———→ Super VHS select
(14) ———→ AV signal select
(15) ←———→ Keypad function scanning
(16) ←———→ Keypad function scanning
(17) ←———→ Keypad function scanning
(18) ←———→ Keypad function scanning
(19) ←———→ Keypad function scanning
(20) ←———→ Keypad function scanning
(21) Chassis
(22) ———→ Red on screen display
(23) ———→ Green on screen display
(24) ———→ Blue on screen display
(25) ———→ On screen display blanking
(26) ←———→ On screen display horizontal sync pulse
(27) ←———→ On screen display vertical sync pulse
(28) ←———→ On screen display oscillator
(29) ←———→ On screen display oscillator
(30) Chassis
(31) ←———→ System oscillator
(32) ←———→ System oscillator
(33) ←——— Power on reset = low
(34) ←——— Received signal ident sync pulse
(35) ←——— Remote control data
(36) ———→ NTSC transmission system select
(37) N/c
(38) N/c
(39) ———→ Serial clock bus
(40) ←———→ Serial data bus
(41) ———→ Power on/standby control
(42) ←——— 5 V supply

Variation

(7), (8), (11), (13) Chassis

Notes:

SAA1250

Infra-red remote controller IC

Pins
(1) Ground
(2) ⟶ Oscillator R/C components (no external xtal.)
(3) ⟶ Oscillator R/C components (no external xtal.)
(4) ⟶ Oscillator R/C components (no external xtal.)
(5) ⟶ Infra-red transmitting diode driver
(6) ⟵ 9 V supply
(7) ⟵ 9 V supply
(8) ⟷ Keypad functions scanning
(9) ⟷ Keypad functions scanning
(10) ⟷ Keypad functions scanning
(11) ⟷ Keypad functions scanning
(12) ⟷ Keypad functions scanning
(13) ⟷ Keypad functions scanning
(14) ⟷ Keypad functions scanning
(15) ⟷ Keypad functions scanning
(16) ⟷ Keypad functions scanning
(17) ⟷ Keypad functions scanning
(18) ⟷ Keypad functions scanning
(19) ⟷ Keypad functions scanning
(20) ⟷ Keypad functions scanning
(21) ⟷ Keypad functions scanning
(22) ⟷ Keypad functions scanning
(23) ⟷ Keypad functions scanning
(24) ⟵ 9 V supply

1 24
12

Variation

(9), (11), (12), (13), (14) N/c

Notes:

SAA1288

Control microprocessor IC

Pins

(1) ←→ System oscillator 4 MHz
(2) ← 5 V supply
(3) → External memory IC
(4) ← Power on reset = low
(5) → Power on/standby control
(6) Chassis
(7) ←→ IM bus data external memory IC
(8) → IM bus clock external memory IC
(9) → IM bus control, data transfer = low, external memory IC
(10) → Picture contrast control
(11) → Picture brightness control
(12) ← Remote control data
(13) → External memory IC mode control
(14) ←→ Keypad functions scanning & seven segment LED display
(15) ←→ Keypad functions scanning & seven segment LED display
(16) ←→ Keypad functions scanning & seven segment LED display
(17) ←→ Keypad functions scanning & seven segment LED display
(18) ←→ Keypad functions scanning & seven segment LED display
(19) ←→ Keypad functions scanning & seven segment LED display
(20) Chassis
(21) → Seven segment LED display
(22) ←→ Keypad functions scanning & seven segment LED display
(23)
(24)
(25) → Seven segment LED display
(26) → Seven segment LED display
(27) ← 5 V supply
(28) ←→ Data bus, tuner & teletext decoder
(29) → Clock bus, tuner & teletext decoder

(Continued opposite)

(30)
(31) ⟶ AV /TV signal source select
(32) ⟶ Teletext sync pulse
(33) ⟶ Picture colour control
(34) ⟶ Sound volume control
(35)
(36) ⟵ Received signal ident sync pulse
(37) ⟵ SCART socket status pin 8
(38) ⟷ Keypad functions scanning
(39) ⟷ Keypad functions scanning
(40) ⟵ 5 V supply

Notes:

SAA1293

Control microprocessor IC

Pins
(1) ⟷ System oscillator 4 MHz
(2) ⟵ 5 V supply
(3) ⟶ Serial clock, external memory IC
(4) ⟵ Power on reset = low
(5) ⟶ Power on/standby control
(6) Chassis
(7) ⟷ IM bus data
(8) ⟶ IM bus control, data transfer = low
(9) ⟶ IM bus clock
(10) ⟶ Picture brightness control
(11) ⟶ Picture colour control
(12) ⟵ Remote control data
(13) ⟶ TV channel tuning control
(14) ⟷ Keypad function scanning & seven segment LED display
(15) ⟷ Keypad function scanning & seven segment LED display
(16) ⟷ Keypad function scanning & seven segment LED display

1 40 20

(Continued overleaf)

SAA1293

(17) ↔ Keypad function scanning & seven segment LED display
(18) ↔ Keypad function scanning & seven segment LED display
(19) ↔ Keypad function scanning & seven segment LED display
(20) Chassis
(21) ↔ Keypad function scanning & seven segment LED display
(22) ↔ Keypad function scanning & seven segment LED display
(23) ↔ Keypad function scanning & seven segment LED display
(24) ↔ Keypad function scanning & seven segment LED display
(25) ↔ ICL data bus
(26) → ICL clock bus
(27) ← 5 V supply
(28)
(29) → TV channel tuning band select, UHF = high
(30) → TV channel tuning band select, UHF = high
(31) → TV/RGB signal source select, TV = low
(32) → TV/AV signal source select
(33) → Picture contrast control
(34) → Sound volume control
(35) → TV channel auto frequency control
(36) ↔ Keypad function scanning
(37) ↔ Keypad function scanning
(38) ↔ Keypad function scanning
(39) ↔ Keypad function scanning
(40) ← 5 V supply

Variation

(5) → Power switch pulse contact
(11) → Contrast picture control
(25), (26) N/c
(33) → Picture colour control

Notes:

SAA1294 - 2

Control microprocessor IC

Pins
(1) ←——→ System oscillator 4 MHz
(2) ←—— 5 V supply
(3) ——→ External memory IC
(4) ←—— Power on reset = low
(5) ——→ Power on/standby control
(6) Chassis
(7) ←——→ IM bus data, external memory IC & circuits
(8) ——→ IM bus control, data transfer = low, external memory IC & circuits
(9) ——→ IM bus clock, external memory IC & circuits
(10) ——→ Sound volume
(11) ——→ Picture colour control
(12) ←—— Remote control data
(13) ——→ TV channel tuning control
(14) ←——→ Keypad function scanning & seven segment LED display
(15) ←——→ Keypad function scanning & seven segment LED display
(16) ——→ Seven segment LED display
(17) ——→ Seven segment LED display
(18) ——→ Seven segment LED display
(19) ←——→ Keypad function scanning & seven segment LED display
(20) Chassis
(21) ←——→ Keypad function scanning & seven segment LED display
(22) ←——→ Keypad function scanning & seven segment LED display
(23) ←——→ Keypad function scanning & seven segment LED display
(24) ←——→ Keypad function scanning & seven segment LED display
(25) ——→ AV signal source select
(26)
(27) ——→ Standby LED indicator
(28) ——→ SECAM/NTSC transmission system select

(Continued overleaf)

SAA1296A

(29)⟶ TV tuning band UHF/ VHF select
(30)⟶ TV tuning band UHF/ VHF select
(31)⟶ SECAM/NTSC transmission system select
(32)⟵ 5 V supply
(33)⟶ Picture contrast control
(34)⟶ Picture brightness control
(35)⟶ TV channel auto frequency control mute
(36)⟷ Keypad function scanning & seven segment LED display
(37)⟷ Keypad function scanning & seven segment LED display
(38)⟷ Keypad function scanning & seven segment LED display
(39)⟷ Keypad function scanning & seven segment LED display
(40)⟵ 5 V supply

Notes:

SAA1296A

Control microprocessor IC

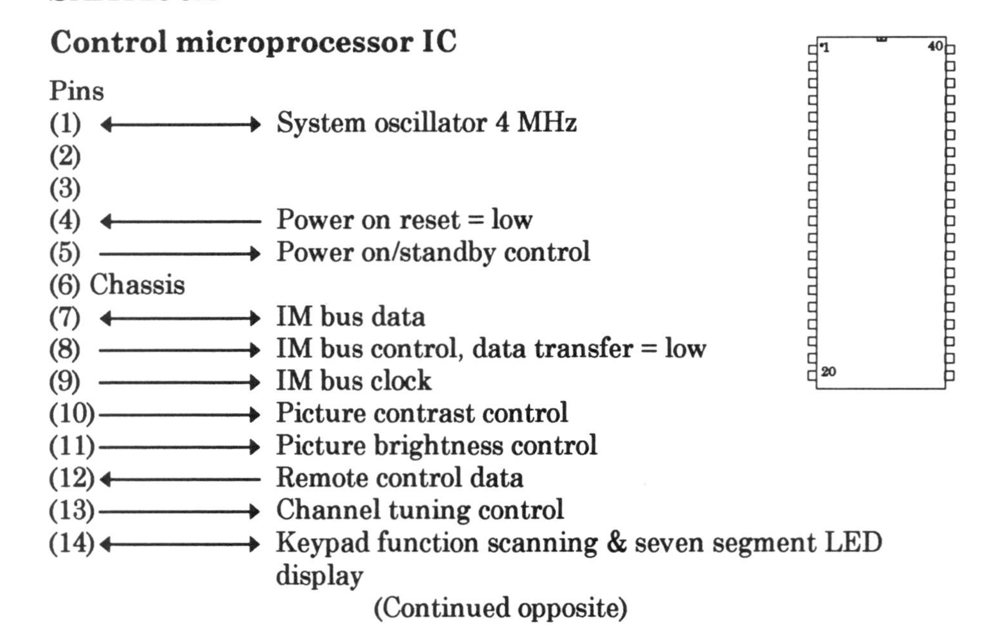

Pins
(1) ⟷ System oscillator 4 MHz
(2)
(3)
(4) ⟵ Power on reset = low
(5) ⟶ Power on/standby control
(6) Chassis
(7) ⟷ IM bus data
(8) ⟶ IM bus control, data transfer = low
(9) ⟶ IM bus clock
(10)⟶ Picture contrast control
(11)⟶ Picture brightness control
(12)⟵ Remote control data
(13)⟶ Channel tuning control
(14)⟷ Keypad function scanning & seven segment LED display

(Continued opposite)

(15) ←——→ Keypad function scanning & seven segment LED display
(16) ←——→ Keypad function scanning & seven segment LED display
(17) ←——→ Keypad function scanning & seven segment LED display
(18) ←——→ Keypad function scanning & seven segment LED display
(19) ←——→ Keypad function scanning & seven segment LED display
(20)
(21) ←——→ Keypad function scanning & seven segment LED display
(22) ←——→ Keypad function scanning & seven segment LED display
(23) ←——→ Keypad function scanning & seven segment LED display
(24) ←—— TV channel auto fine tuning
(25)
(26)
(27) ←—— 5 V supply
(28) ←——→ Keypad function scanning & seven segment LED display
(29)
(30)
(31)
(32) ——→ TV/AV signal source select
(33) ——→ Picture colour control
(34) ——→ Sound volume control
(35)
(36) ←—— SCART socket status pin 8
(37)
(38) ←——→ Keypad function scanning & seven segment LED display
(39) ←——→ Keypad function scanning & seven segment LED display
(40) ←—— 5 V supply

Notes:

SAA1298

Control microprocessor IC

Pins
(1) ←——— System oscillator 4 MHz
(2) ←——— 5 V supply
(3) ———→ External memory IC mode control
(4) ←——— Power on reset = low
(5) ———→ Power on/standby control
(6) Chassis
(7) ←——→ IM bus data
(8) ———→ IM bus control
(9) ———→ IM bus clock
(10) ———→ Sound volume control
(11) ———→ Picture colour control
(12) ←——— Remote control data
(13) ———→ TV channel tuning control
(14) ←——→ Keypad function scanning & seven segment LED display
(15) ←——→ Keypad function scanning & seven segment LED display
(16) ←——→ Keypad function scanning & seven segment LED display
(17) ←——→ Keypad function scanning & seven segment LED display
(18) ←——→ Keypad function scanning & seven segment LED display
(19) ←——→ Keypad function scanning & seven segment LED display
(20) Chassis
(21) ←——→ Keypad function scanning & seven segment LED display
(22) ←——→ Keypad function scanning & seven segment LED display
(23) ———→ Seven segment LED display supply
(24) ———→ Seven segment LED display supply
(25) ———→ TV/AV signal source select
(26) ←——→ Data, ICL DLIM data bus (data limited clock)
(27) ←——— 5 V supply
(28) ———→ Video & sound switching
(29) ———→ TV tuning band VHF/UHF select

(Continued opposite)

(30) ⟶ TV tuning band VHF/UHF select
(31) ⟶ AV/RGB signal source select
(32) ⟵ 5 V supply
(33) ⟶ Picture contrast control
(34) ⟶ Picture brightness control
(35) ⟶ TV channel auto frequency control
(36) ⟷ Keypad function scanning
(37) ⟷ Keypad function scanning
(38) ⟷ Keypad function scanning
(39) ⟷ Keypad function scanning
(40) ⟵ 5 V supply

Notes:

SAA3004

Infra-red remote controller IC

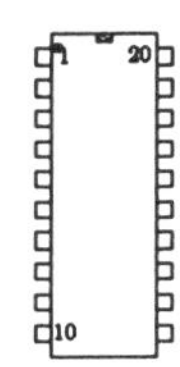

Pins
(1) ⟶ Infra-red transmitting diode driver
(2) N/c
(3) ⟷ Keypad functions scanning
(4) ⟷ Keypad functions scanning
(5) ⟷ Keypad functions scanning
(6) ⟷ Keypad functions scanning
(7) ⟷ Keypad functions scanning
(8) ⟷ Keypad functions scanning
(9) ⟷ Keypad functions scanning
(10) Ground
(11) ⟷ System oscillator 455 kHz
(12) ⟷ System oscillator455 kHz
(13) ⟷ Keypad functions scanning
(14) ⟷ Keypad functions scanning
(15) ⟷ Keypad functions scanning
(16) ⟷ Keypad functions scanning
(17) ⟷ Keypad functions scanning

(Continued overleaf)

(18) ←——→ Keypad functions scanning
(19) ←——→ Keypad functions scanning
(20) ←——→ 6 V supply

Variations

(2) ←——→ Keypad functions scanning

Notes:

SAA3006T

Infra-red remote controller IC

Pins
(1) ←——→ Keypad functions scanning
(2) ←—— 4V5 supply
(3) ——→ TV/ VCR select
(4) N/c
(5) N/c
(6) N/c
(7) ——→ Infra-red transmitting diode driver
(8) N/c
(9) ←——→ Keypad functions scanning
(10) ←——→ Keypad functions scanning
(11) ←——→ Keypad functions scanning & pin 3
(12) ←——→ Keypad functions scanning
(13) ←——→ Keypad functions scanning
(14) Ground
(15) ←——→ Keypad functions scanning
(16) ←——→ Keypad functions scanning
(17) ←——→ Keypad functions scanning
(18) ←——→ System oscillator 429 kHz
(19) Ground
(20) Ground
(21) ←——→ Keypad functions scanning
(22) ←——→ Keypad functions scanning
(23) ←——→ Keypad functions scanning
(24) ←——→ Keypad functions scanning

(Continued opposite)

(25) ←——→ Keypad functions scanning
(26) ←——→ Keypad functions scanning
(27) ←——→ Keypad functions scanning
(28) ←—— 4V5 supply

Notes:

SAA3008

Infra-red remote controller IC

Pins
(1) ——→ Infra-red transmitting diode driver
(2) N/c
(3) N/c
(4) N/c
(5) N/c
(6) ←——→ Keypad functions scanning
(7) ←——→ Keypad functions scanning
(8) ←——→ Keypad functions scanning
(9) N/c
(10) Ground
(11) ←——→ System oscillator 455KHz
(12) ←——→ System oscillator 455KHz
(13) ←——→ Keypad functions scanning
(14) ←——→ Keypad functions scanning
(15) ←——→ Keypad functions scanning
(16) N/c
(17) ←——→ Keypad functions scanning
(18) ←——→ Keypad functions scanning
(19) ←——→ Keypad functions scanning
(20) ←—— 3 V supply

Notes:

SAA3010T

Infra-red remote controller IC

Pin
(1) ←——→ Keypad functions scanning
(2) ←—— 3 V or 6 V Supply
(3) ←——→ Keypad functions scanning
(4) N/c
(5) N/c
(6) N/c
(7) ——→ Infra-red transmitting diode driver
(8) N/c
(9) ←——→ Keypad functions scanning
(10) ←——→ Keypad functions scanning
(11) ←——→ Keypad functions scanning
(12) ←——→ Keypad functions scanning
(13) ←——→ Keypad functions scanning
(14) Ground
(15) ←——→ Keypad functions scanning
(16) ←——→ Keypad functions scanning
(17) ←——→ Keypad functions scanning
(18) ←——→ System oscillator 455 kHz
(19) Ground
(20) Ground
(21) ←——→ Keypad functions scanning
(22) ←——→ Keypad functions scanning
(23) ←——→ Keypad functions scanning
(24) ←——→ Keypad functions scanning
(25) ←——→ Keypad functions scanning
(26) ←——→ Keypad functions scanning
(27) ←——→ Keypad functions scanning
(28) ←—— 3 V or 6 V Supply

Variation
(24) N/c

Notes:

SAA3027

Infra-red remote controller IC

Pins
(1) ←———→ Keypad functions scanning
(2) ←——— 9 V supply
(3) ←———→ Keypad functions scanning
(4) N/c
(5) ←——— 9 V supply
(6) N/c
(7) ———→ Infra-red transmitting diode driver
(8) N/c
(9) ←———→ Keypad functions scanning
(10) ←———→ Keypad functions scanning
(11) ←———→ Keypad functions scanning
(12) ←———→ Keypad functions scanning
(13) ←———→ Keypad functions scanning
(14) Ground
(15) ←———→ Keypad functions scanning
(16) ←———→ Keypad functions scanning
(17) ←———→ Keypad functions scanning
(18) ←———→ Oscillator (no system)
(19) Ground
(20) ←———→ Oscillator (no system)
(21) ←———→ Keypad functions scanning
(22) ←———→ Keypad functions scanning
(23) ←———→ Keypad functions scanning
(24) ←———→ Keypad functions scanning
(25) ←———→ Keypad functions scanning
(26) N/c
(27) N/c
(28) ←——— 9 V supply

Notes:

SAA5246A

SAA5246A

Control microprocessor IC

Pins
(1) ←——— 5 V supply
(2)
(3) ←———→ System oscillator 27 MHz
(4) ←———→ System oscillator 27 MHz
(5) Chassis
(6)
(7)
(8) ←——— Received signal ident sync pulse
(9)
(10) ←——— 5 V supply
(11) Chassis
(12) ←——— On screen display horizontal sync pulse
(13) ←——— On screen display vertical sync pulse
(14) Chassis
(15) ———→ Red on screen display
(16) ———→ Green on screen display
(17) ———→ Blue on screen display
(18) ←——— 5 V supply
(19) ←——— RGB SCART socket fast blanking pin 16
(20) N/c
(21) ←——— Vertical flyback pulse Teletext mode
(22) N/c
(23) ←———→ IIC clock bus
(24) ←———→ IIC data bus
(25) Chassis
(26) ←———→ External memory IC data
(27) ←———→ External memory IC data
(28) ←———→ External memory IC data
(29) ←———→ External memory IC data
(30) ←———→ External memory IC data
(31) ←———→ External memory IC data
(32) ←———→ External memory IC data
(33) ←———→ External memory IC data
(34) ———→ External memory IC address
(35) ———→ External memory IC address
(36) ———→ External memory IC address
(37) ———→ External memory IC address

(Continued opposite)

(38) ——→ External memory IC address
(39) ——→ External memory IC address
(40) ——→ External memory IC address
(41) ——→ External memory IC address
(42) ——→ External memory IC address
(43) ——→ External memory IC address
(44) ——→ External memory IC address
(45) ——→ External memory IC address
(46) ——→ External memory IC address
(47) ——→ External memory IC enable = low
(48) ——→ External memory read/write mode control

Notes:

SAB2083

Control microprocessor IC

Pins
(1) N/c
(2) N/c
(3) ——→ NICAM decoder enable
(4) ——→ NICAM mono/stereo, bilingual select
(5) ——→ NICAM mono/stereo, bilingual select
(6) ——→ Received signal auto fine tuning
(7) ——→ Received signal auto fine tuning
(8) ——→ NICAM function LED indicator
(9) ——→ Standby LED indicator , on = low
(10) Chassis
(11) ←—— 5 V supply
(12) ←——→ System oscillator 12 MHz
(13) ←——→ System oscillator 12 MHz
(14) ←—— Power on reset = low
(15) ——→ NTSC transmission, picture hue control
(16) ←——→ Keypad function scanning
(17) ←——→ Keypad function scanning
(18) ←——→ Keypad function scanning
(19) N/c
(20) ←—— Scanning/video system detect

(Continued overleaf)

(21)———→ VCR signal source select
(22)———→ Sound mute
(23)←——— 50 Hz/60 Hz scan system detect, 50 Hz = low
(24)←——— Vertical sync pulse
(25)←——— Remote control data
(26)←——→ Data bus, teletext circuit
(27)———→ Clock bus, teletext circuit
(28)←——— Received signal ident sync pulse
(29)←——— Power on/standby control
(30)———→ Picture blanking, text/RGB processing circuits
(31)←——→ Data bus, external memory IC/on screen display IC/tuner, etc.
(32)———→ Clock bus
(33) N/c
(34)———→ SCART socket RGB blanking status pin 16
(35)———→ Sound signal input source switching ICs
(36)←——— SCART socket 1 status pin 8
(37)←——— SCART socket 2 status pin 8
(38)———→ Luminance/colour video signal switching ICs
(39) N/c
(40) N/c

Notes:

SAB3034

Control microprocessor IC

Pins
(1) Chassis
(2)
(3) ←——— Clock reference 400 kHz
(4) ———→ Sound volume control
(5) ———→ Picture colour control
(6) ———→ Picture brightness control
(7) ———→ Sound bass control
(8) ———→ Sound treble control
(9) ———→ Tuning auto frequency control
(10)←——— 5 V supply

(Continued opposite)

(11) ← Clock bus
(12) ← Bus control/enable
(13) ← Data bus
(14) → Tuning auto frequency control on/off control
(15) → TV channel tuning down
(16) → TV channel tuning up
(17) Chassis
(18) ← Tuner oscillator sample, 64, pre-scaler

Notes:

SAB3035

Control microprocessor IC

Pins
(1) → Sound volume control
(2) N/c
(3) N/c
(4) → Picture colour control
(5) ↔ Serial data bus
(6) → Serial clock bus
(7) N/c
(8) N/c
(9) N/c
(10) ← Received signal ident sync pulse(tuning search stop)
(11) ← TV channel auto frequency control up
(12) ← TV channel auto frequency control down
(13)
(14) Chassis
(15) → TV channel tuning control
(16) ← 12 V supply
(17) ← 33 V supply
(18) → TV tuning band UHF
(19) → TV tuning band VHF H
(20) → TV tuning band VHF L
(21) → TV tuning band hyperband
(22) ← 12 V supply
(23) ← Tuner oscillator sample (pre-scaler)

(Continued overleaf)

(24) ←——→ Reference oscillator 4 MHz
(25) ——→ Picture contrast control
(26) ——→ Picture brightness control
(27) N/c
(28) N/c

Variation

(1) ——→ Sound mute
(2) ——→ Sound balance
(3) ——→ Sound bass
(4) ——→ Sound treble
(5) ←——→ Serial data bus
(6) ——→ Serial clock bus
(7) ——→ AV signal source switching
(8) Chassis
(9) ←—— Received TV signal ident sync pulse
(10) N/c
(25) ——→ Picture brightness control
(26) ——→ Picture contrast control
(27) ——→ Picture colour control
(28) ——→ Sound volume control

Variation

(21) ——→ AV/TV signal source select
(27) ——→ Sound treble control
(28) ——→ Sound bass

Variation

(1) N/c
(2) ——→ Sound mute
(3) ——→ Teletext enable
(4) ——→ TV / AV signal select
(9) ←—— SCART socket status pin 8

Notes:

SAB3036

Control microprocessor IC

Pins
(1) Chassis
(2) Chassis
(3) Chassis
(4) Chassis
(5) ←——— 5 V supply
(6)
(7) Chassis
(8) ———→ TV channel tuning
(9) ←——— 33 V supply
(10) N/c
(11) N/c
(12) N/c
(13) ———→ TV tuning band select, UHF = low
(14) ←——— 12 V supply
(15) ←——— Tuner oscillator sample (pre-scaler)
(16) ←———→ System oscillator
(17) ←———→ Serial data bus
(18) ←——— Serial clock bus

Notes:

SDA2023

Control microprocessor IC

Pins
(1) N/c
(2) ←———→ Keypad function scanning & seven segment LED display
(3) ←———→ Keypad function scanning & seven segment LED display
(4) ←———→ Keypad function scanning & seven segment LED display

(Continued overleaf)

SDA2023

(5) ←→ Keypad function scanning & seven seg. LED display

(6) ←→ Keypad function scanning & seven seg. LED display

(7) ←→ Keypad function scanning & seven seg. LED display

(8) ←→ Keypad function scanning & seven seg. LED display

(9) ←→ Keypad function scanning & seven seg. LED display

(10) Chassis

(11) ← 5 V supply

(12) ←→ System oscillator 12 MHz

(13) ←→ System oscillator 12 MHz

(14) ← Power on reset = low

(15) → Sound volume control

(16) → Picture brightness control

(17) → Picture colour control

(18) → Picture contrast control

(19) → NTSC transmission picture hue control

(20) → Power on/standby control

(21) ← 50 Hz/60 Hz scan system detect

(22) ← SCART socket status pin 8

(23) → Teletext circuit serial clock

(24) ←→ Teletext circuit serial data

(25)

(26) → Teletext circuit, ICL clock bus,(data limited clock)

(27) ← Remote control data

(28) → Text enable

(29) ← Circuit protection monitoring

(30)

(31) ←→ Serial data bus

(32) → Serial clock bus

(33) ←→ Keypad function scanning & seven seg. LED display

(34) ←→ Keypad function scanning & seven seg. LED display

(35) ←→ Keypad function scanning & seven seg. LED display

(36) ←→ Keypad function scanning

(37) ←→ Keypad function scanning

(38) ←→ Keypad function scanning

(39) ←→ Keypad function scanning

(40) → Power switch pulse contact

Notes:

SDA2080

Control microprocessor IC

Pins

(1) ⟶ Inverted RGB select
(2) ⟶ Sound circuit supply on/off
(3) ⟶ Picture circuit supply on /off
(4) N/c
(5) N/c
(6)
(7)
(8)
(9)
(10) ⟵ SCART socket 1 status pin 8
(11) ⟵ SCART socket 2 status pin 8
(12) ⟵ Remote control data
(13) N/c
(14) ⟷ Keypad function scanning & power switch pulse contact
(15) ⟷ Keypad function scanning
(16) ⟷ Keypad function scanning
(17) ⟷ Keypad function scanning
(18) ⟷ System oscillator 8.8 MHz
(19) ⟷ System oscillator 8.8 MHz
(20) Chassis
(21) ⟷ Serial data bus
(22) ⟶ Serial clock bus
(23) ⟷ Keypad function scanning
(24) ⟷ Keypad function scanning power switch pulse contact
(25) ⟷ Keypad function scanning
(26) N/c
(27) ⟷ Keypad function scanning
(28) ⟶ LED indicator
(29) ⟶ LED indicator
(30) ⟶ LED indicator
(31) N/c
(32) ⟶ Seven segment LED display
(33) ⟶ Seven segment LED display
(34) ⟶ Seven segment LED display
(35) ⟶ Seven segment LED display
(36) ⟶ Seven segment LED display

(Continued overleaf)

SDA2084 - A002

(37) ⟶ Seven segment LED display
(38) ⟶ Seven segment LED display
(39) ⟶ Seven segment LED display
(40) ⟵ 5 V supply

Notes:

SDA2084 - A002

Control microprocessor IC

Pins
(1) N/c
(2) ⟶ Picture brightness control
(3) ⟶ Picture contrast control
(4) ⟶ Picture colour control
(5) ⟶ Seven segment LED display
(6) ⟶ Seven segment LED display
(7) N/c
(8) ⟵ 5 V supply
(9) ⟵ Power on reset = low
(10) ⟵ Remote control data
(11) ⟶ AV select
(12) ⟶ Linked to pin 10
(13) N/c
(14) N/c
(15) ⟵ Received signal ident sync pulse
(16) N/c
(17) ⟶ Power on/standby control
(18) ⟷ System oscillator 12 MHz
(19) ⟷ System oscillator 12 MHz
(20) Chassis
(21) ⟷ Serial data bus
(22) ⟶ Serial clock bus
(23) ⟵ 5 V supply
(24) ⟵ 5 V supply
(25) Chassis
(26) Chassis
(27) N/c

1 40 20

(Continued opposite)

(28) N/c
(29) ←——→ Keypad function scanning
(30) ←——→ Power switch pulse contact
(31) N/c
(32) ←——→ Keypad function scanning & seven seg. LED display
(33) ←——→ Keypad function scanning & seven seg. LED display
(34) ←——→ Keypad function scanning & seven seg. LED display
(35) ←——→ Keypad function scanning & seven seg. LED display
(36) ←——→ Keypad function scanning & seven seg. LED display
(37) ←——→ Keypad function scanning & seven seg. LED display
(38) ←——→ Keypad function scanning & seven seg. LED display
(39) ←——→ Keypad function scanning & seven seg. LED display
(40) ←—— 5 V supply

Notes:

SDA2208

Infra-red remote controller IC

Pins
(1) Ground
(2) ——→ Infra-red transmitting diode driver
(3) ←——→ Keypad functions scanning
(4) ←——→ Keypad functions scanning
(5) ←——→ Keypad functions scanning
(6) ←——→ Keypad functions scanning
(7) ←——→ Keypad functions scanning
(8) ←——→ Keypad functions scanning
(9) ←——→ Keypad functions scanning
(10) ←——→ Keypad functions scanning
(11) N/c
(12) ←——→ Keypad functions scanning
(13) ←——→ Keypad functions scanning
(14) ←——→ Keypad functions scanning
(15) ←——→ Keypad functions scanning

(Continued overleaf)

(16)←——→ Keypad functions scanning
(17)←——→ Keypad functions scanning
(18)←——→ Keypad functions scanning
(19)←——→ Keypad functions scanning
(20)←——→ System oscillator 485 kHz

Notes:

SDA3208

Infra-red remote controller IC

Pins
(1) ←——→ Keypad functions scanning
(2) ←——→ Keypad functions scanning
(3) ←——→ Keypad functions scanning
(4) ←——→ Keypad function scanning
(5) ←——→ Keypad function scanning
(6) N/c
(7) N/c
(8) N/c
(9) ←——→ Keypad functions scanning
(10)←——→ Keypad functions scanning
(11)←——→ Keypad functions scanning
(12)←——→ Keypad functions scanning
(13)←——→ Keypad functions scanning
(14)←——→ Keypad functions scanning
(15)←——→ Keypad functions scanning
(16)←——→ Keypad functions scanning
(17)←——→ Keypad functions scanning
(18) N/c
(19) N/c
(20)←——→ Keypad function scanning
(21)←—— 4 V5 supply
(22)←——→ System oscillator 500 kHz
(23) N/c

(Continued opposite)

(24)———→ Infra-red transmitting diode driver
(25) Ground
(26)←——— 4 V5 supply
(27)←——→ Keypad functions scanning
(28)←——→ Keypad functions scanning

Notes:

SDA20320

Control microprocessor IC

Pins
(1) N/c
(2) ←——— Power rail sensing, (settings memorized)
(3) ←——— Headphone inserted detection
(4) ———→ NTSC/PAL external luminance delay line IC
(5) ———→ NTSC/PAL external luminance delay line IC
(6) ———→ LED indicator
(7) N/c
(8) ———→ Text source switching IC
(9) ———→ On screen display source switching
(10) Chassis
(11)———→ Power on/standby control
(12)←——→ System oscillator 8 MHz
(13)←——→ System oscillator 8 MHz
(14)←——— Power on reset = low
(15)———→ Stereo crosstalk control
(16) N/c
(17) N/c
(18)———→ Vertical scan shift control
(19)———→ Tuner auto gain control
(20) N/c
(21)———→ Sound beep IC
(22) N/c
(23)———→ External satellite power supply on /off control

(Continued overleaf)

(24)——————→ RGB tube drive IC on/off switching
(25)←—————— Remote control data
(26)——————→ Multiple AV signal source switching control IC
(27)——————→ Multiple AV signal source switching control IC
(28)——————→ Multiple AV signal source switching control IC
(29) N/c
(30)←—————— Vertical sync pulse teletext
(31) N/c
(32) N/c
(33)←—————→ Serial data bus
(34)——————→ Serial clock bus
(35)←—————→ Keypad function scanning
(36)←—————— SCART socket 2 status pin 8
(37)←—————— SCART socket 1 status pin 8
(38)←—————— Received signal ident sync pulse
(39)←—————→ Keypad function scanning
(40)←—————— SCART socket RGB status pin 16

Notes:

SDA20560

Control microprocessor IC

Pins
(1) N/c
(2) ——————→ Picture blanking
(3) N/c
(4) ——————→ Standby LED indicator on = low
(5) ——————→ TV signal auto fine tuning down
(6) ——————→ TV signal auto fine tuning up
(7) ——————→ RGB data, external on screen display IC
(8) ——————→ RGB data, external on screen display IC
(9) ——————→ RGB data, external on screen display IC
(10) Chassis
(11)←—————— 5 V supply
(12)←—————→ System oscillator 10 MHz
(13)←—————→ System oscillator 10 MHz
(14)←—————— Power on reset = low
(15)——————→ NTSC transmission, picture hue control

(Continued opposite)

(16)———→ SCART socket RGB blanking status pin 16
(17)———→ Transmission system, PAL/SECAM/NTSC switching
(18)———→ Luminance/colour signal source select
(19)←——— SCART socket 1 status pin 8
(20)←——— SCART socket 2 status pin 8
(21)———→ VTR mode, timebase time constant enable
(22)←——→ Keypad function scanning
(23)←——→ Keypad function scanning
(24)←——→ Keypad function scanning
(25)←——— Remote control data
(26)←——→ Teletext circuit data bus
(27)———→ Teletext circuit clock bus
(28)←——— Received signal ident sync pulse
(29)———→ Power on/standby control
(30)———→ Sound mute
(31)←——→ Serial data bus
(32)———→ Serial clock bus
(33)←——— On screen display horizontal sync pulse
(34)←——— On screen display vertical sync pulse
(35)←——— External on screen display IC data
(36) N/c
(37)———→ Text character, video processing circuit
(38)———→ On screen display picture blanking
(39)
(40) N/c

Variation

(22), (23), (24), (38) N/c

Notes:

ST6326

ST6326

Control microprocessor IC

Pins
(1) N/c
(2) N/c
(3) N/c
(4) N/c
(5) N/c
(6) Chassis
(7) Chassis
(8) ←——→ On screen display oscillator
(9) ←——→ On screen display oscillator
(10) N/c
(11) ←——→ Serial data bus
(12) ——→ Serial clock bus
(13) ←—— On screen display horizontal sync pulse
(14) ←—— On screen display vertical sync pulse
(15) N/c
(16) N/c
(17) ←—— Beam current protection sensing
(18) ——→ Power on/standby control
(19) ←—— 50 Hz /60 Hz system scanning detect
(20) Chassis
(21) ——→ Power switch pulse contact
(22) N/c
(23) N/c
(24) ←—— Power on reset = low
(25) ←——→ System oscillator 8 MHz
(26) ←——→ System oscillator 8 MHz
(27) ←—— Remote control data
(28) N/c
(29) N/c
(30) ——→ RGB text/SECAM/NTSC/PAL signal select IC
(31) ——→ Red on screen display
(32) ——→ Green on screen display
(33) ——→ Blue on screen display
(34)
(35) N/c
(36) ——→ Picture brightness control
(37) ——→ Picture colour control

(Continued opposite)

(38) ——→ Picture contrast control
(39) ——→ Sound volume control
(40) ←—— 5 V supply

Notes:

ST6393

Control microprocessor IC

Pins
(1) ——→ Picture aspect ratio select, low = 16:9, high = 4:3
(2) ——→ Picture contrast control
(3) ——→ Picture colour control
(4) ——→ Picture brightness control
(5) ——→ Sound volume control
(6)
(7) ——→ RGB mode select (if SCART socket pin 16 blanking not present)
(8) ←——→ Keypad function scanning
(9) ←—— SCART socket status pin 8 (5 V = 16:9, 12 V = 4:3 picture aspect ratio)
(10) ——→ TV enable = high (if SCART socket RGB plus blanking present)
(11) ←——→ Keypad function scanning
(12) ←——→ Keypad function scanning
(13) ←——→ Keypad function scanning
(14) ←——→ Keypad function scanning
(15) ←——→ Keypad function scanning
(16) ←——→ Keypad function scanning
(17) ——→ TV/AV signal select
(18) ——→ Super VHS = high
(19) ——→ LED indicator pulse during child lock mode
(20) ——→ Power on/standby control
(21) Chassis
(22) ——→ Red on screen display
(23) ——→ Green on screen display
(24) ——→ Blue on screen display

(Continued overleaf)

(25) ——→ On screen display picture blanking

ST6393

(25) ——→ On screen display picture blanking
(26) ←—— On screen display horizontal sync pulse
(27) ←—— On screen display vertical sync pulse
(28) ——→ On screen display oscillator
(29) ——→ On screen display oscillator
(30)
(31) ←——→ System oscillator 8 MHz
(32) ←——→ System oscillator 8 MHz
(33) ←—— Power on reset = low
(34) ←—— TV settings memorized at power loss
(35) ←—— Remote control data
(36) ←—— Search tuning stop, signal received ident sync pulse
(37) ——→ Tuner PAL L = low, PAL B/G = high select
(38) N/c
(39) N/c
(40) ←——→ 12C data bus, tuner & teletext circuits
(41) ——→ 12C clock bus, tuner & teletext circuits
(42) ←—— 5 V supply

Notes:

ST6393B1

Control microprocessor IC

Pins
(1) N/c
(2) ⟶ Tuner auto gain control
(3) ⟶ Sound mute = high
(4)
(5) ⟶ RGB source switching IC
(6) N/c
(7) ⟷ Volume down keypad function scanning
(8) ⟷ TV channel down keypad function scanning
(9) ⟶ TV channel auto frequency control up
(10) ⟷ Volume up keypad function scanning
(11) ⟶ Received signal ident sync pulse
(12) ⟶ TV channel auto frequency control down
(13) ⟶ Power on/standby control
(14) N/c
(15) N/c
(16) Chassis
(17)
(18) ⟶ Low = LED standby indicator driver
(19) ⟶ Low = LED mono/stereo indicator driver
(20) ⟶ Low = LED mono/stereo indicator driver
(21) Chassis
(22) ⟶ Red on screen display
(23) ⟶ Green on screen display
(24) ⟶ Blue on screen display
(25) ⟶ On screen display picture blanking
(26) ⟵ On screen display horizontal sync pulse
(27) ⟵ On screen display vertical sync pulse
(28) ⟷ On screen display oscillator
(29) ⟷ On screen display oscillator
(30) N/c
(31) ⟷ System oscillator 8 MHz
(32) ⟷ System oscillator 8 MHz
(33) ⟵ Power on reset = low
(34) ⟵ SCART socket 1 status pin 8
(35) ⟵ Remote control data
(36) N/c

(Continued overleaf)

ST6399B1 - KU

(37) ←—— SCART socket 3 status pin 8
(38) ←—— SCART socket 2 status pin 8
(39) ——→ TV channel up keypad function scanning
(40) ←——→ Serial data bus
(41) ——→ Serial clock bus
(42) ←—— 5 V supply

Variation

Pins
(2) N/c
(4) ——→ AV signal source select
(36) ——→ Super VHS select

Notes:

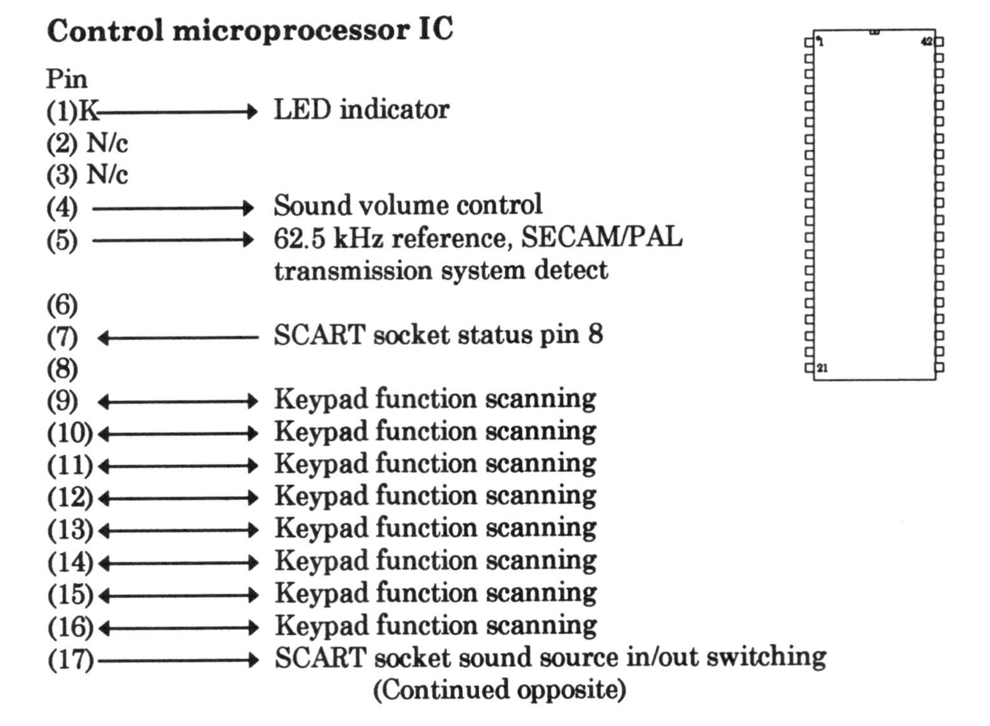

ST6399B1 - KU

Control microprocessor IC

Pin
(1)K ——→ LED indicator
(2) N/c
(3) N/c
(4) ——→ Sound volume control
(5) ——→ 62.5 kHz reference, SECAM/PAL transmission system detect
(6)
(7) ←—— SCART socket status pin 8
(8)
(9) ←——→ Keypad function scanning
(10) ←——→ Keypad function scanning
(11) ←——→ Keypad function scanning
(12) ←——→ Keypad function scanning
(13) ←——→ Keypad function scanning
(14) ←——→ Keypad function scanning
(15) ←——→ Keypad function scanning
(16) ←——→ Keypad function scanning
(17) ——→ SCART socket sound source in/out switching

(Continued opposite)

(18) → Radio alarm circuit
(19) → FM radio on/off
(20) → Power on/standby control
(21) Chassis
(22) → Red on screen display
(23) → Green on screen display
(24) → Blue on screen display
(25) → On screen display picture blanking
(26) ← On screen display horizontal sync pulse
(27) ← On screen display vertical sync pulse
(28) ↔ On screen display oscillator
(29) ↔ On screen display oscillator
(30)
(31) ↔ System oscillator 8 MHz
(32) ↔ System oscillator 8 MHz
(33) ← Power on reset = low
(34) → PAL/SECAM/Transmission system select
(35) ← Remote control data
(36) ← Received signal ident sync pulse
(37) ← Supply monitoring, settings auto memory
(38) → Serial clock bus } Signal source switching & colour decoder
(39) → IC enable/ control } Signal source switching & colour decoder
(40) ↔ Serial data } Tuner, Teletext decoder
(41) → Serial clock } Tuner, Teletext decoder
(42) ← 5 V supply

Notes:

ST9093

Control microprocessor IC

Pins
(1) → Serial 12C clock bus
(2) ↔ Serial 12C data bus
(3) → Teletext 12C clock bus
(4) ↔ Teletext 12C data bus
(5) ← 13 V supply detector OK, initialises external bus connected ICs
(6) ↔ Keypad function scanning
(7) ↔ Keypad function scanning

(Continued overleaf)

(8) ←——→ Keypad function scanning
(9) ←——→ Keypad function scanning

Signal source select

(10) | ——→ Tuner = low, external composite signal = low,
(11) | Tuner = low, external composite signal = high
(10) | ——→ AV1 = high, AV2 = high
(11) | AV1 = low, AV2 = high
(12) ——→ Transmission system select, low = B/G, DKK, I (UK) high = L, PAL-plus ??
(13) ←—— 5 V supply
(14) ——→ Sound volume control
(15) ——→ Satellite standby control
(16) ——→ Picture zoom size select, low = normal, pulsed = 25 % increase high = 33 % increase
(17) ——→ On screen display picture blanking
(18) ——→ Blue on screen display
(19) ——→ Green on screen display
(20) ——→ Red on screen display
(21) Chassis
(22) ←——→ On screen display oscillator
(23) ←——→ On screen display oscillator
(24) ←—— 5 V supply
(25) ←—— On screen display horizontal sync pulse
(26) ←—— On screen display vertical sync pulse
(27) ←—— Auto picture contrast (ambient light sensing)
(28) ←—— SCART socket 1 status pin 8. low = TV, mid = AV 19 : 9, high = AV 4 : 3
(29) ←—— SCART socket 2 status pin 8. low = TV, mid = AV 19 : 9, high = AV 4 : 3
(30) ←——→ System oscillator 11 MHz
(31) Chassis
(32) ←——→ System oscillator 11 MHz
(33) ——→ Power on reset pulse = low
(34) ——→ NTSC select
(35) ——→ LED indicator, high = red, low = green , child lock mode = pulsing
(36) ←—— Remote control data
(37) ——→ Sound mute
(38) ——→ Supply failure settings memorized
(39) ←—— Keypad function scanning
(40) ←—— Teletext decoder data

(Continued opposite)

Variation

Pins
(14) ⟶ External degaussing circuit control

Notes:

TC9012F

Infra-red remote controller IC

Pins
(1) ⟷ Keypad functions scanning
(2) ⟷ Keypad functions scanning
(3) ⟷ Keypad functions scanning
(4) ⟷ Keypad functions scanning
(5) ⟶ Infra-red transmitting diode driver
(6) ⟵ 6 V supply
(7) N/c
(8) ⟷ System oscillator 485kHz
(9) ⟷ System oscillator 485kHz
(10) Ground
(11) N/c
(12) N/c
(13) ⟷ Keypad functions scanning
(14) ⟷ Keypad functions scanning
(15) ⟷ Keypad functions scanning
(16) ⟷ Keypad functions scanning
(17) ⟷ Keypad functions scanning
(18) ⟷ Keypad functions scanning
(19) ⟷ Keypad functions scanning
(20) ⟷ Keypad functions scanning

1 20 10

Notes:

TMP47C433AN

TMP47C433AN

Control microprocessor IC

Pins

(1) ——————→ TV channel tuning control
(2) ——————→ Sound volume control
(3) N/c
(4) N/c
(5) ——————→ TV tuning band VHF L select
(6) ——————→ TV tuning band VHF H select
(7) ——————→ TV tuning band UHF select
(8) ——————→ Power on/standby control
(9) ——————→ AV/ TV select
(10) N/c
(11)——————→ Picture mute
(12)——————→ On screen display picture blanking
(13)←——————→ Data, external on screen display & memory IC
(14)——————→ Clock, external on screen display & memory IC
(15)←——————→ Keypad function scanning & external memory IC
(16)←——————→ Keypad function scanning & external memory IC
(17)←——————→ External memory IC
(18)←——————→ External memory IC
(19)——————→ Sleep LED indicator
(20)←——————→ Keypad function scanning
(21) Chassis
(22)←——————→ Keypad function scanning
(23)←——————→ Keypad function scanning
(24)←——————→ Keypad function scanning
(25)←——————→ Keypad function scanning & power switch pulse contact
(26)←——————→ Keypad function scanning
(27)←——————→ Keypad function scanning
(28)←——————→ Keypad function scanning
(29)←—————— Power switch pulse contact
(30) Chassis
(31)←——————→ System oscillator 4 MHz
(32)←——————→ System oscillator 4 MHz
(33)←—————— Power on reset = low
(34)←—————— Vertical sync pulse & to external on screen display IC
(35)←—————— Remote control data
(36)←—————— Received signal ident sync pulse

(Continued opposite)

TMP47C634NR - 331

Pins		Function
[1]	⟶	Picture sharpness control
[2]	⟶	Picture contrast control
[3]	⟶	Picture colour control
[4]	⟶	Picture brightness control
[5]	⟶	Sound volume control
[6]		
[7]		
[8]	⟷	Keypad function scanning
[9]		
[10]		
[11]	⟷	Keypad function scanning
[12]	⟷	Keypad function scanning
[13]	⟷	Keypad function scanning
[14]	⟷	Keypad function scanning
[15]	⟷	Keypad function scanning
[16]	⟷	Keypad function scanning
[17]		
[18]		
[19]		
[20]	⟶	Power on / standby control = Low
[21]		Chassis
[22]	⟶	Red on screen display
[23]	⟶	Green on screen display
[24]	⟶	Blue on screen display
[25]	⟶	On screen display picture blanking
[26]	⟵	On screen display horizontal sync' pulse
[27]	⟵	On screen display vertical sync' pulse
[28]	⟷	On screen display oscillator
[29]	⟷	On screen display oscillator
[30]		
[31]	⟷	System oscillator 4 Mhz
[32]	⟷	System oscillator 4 Mhz
[33]	⟵	Power on reset = Low
[34]	⟵	Supply failure detect, settings memorized
[35]	⟵	Remote control data
[36]	⟵	Received signal ident sync' pulse
[37]	⟶	Sync' seperator circuit mute (during switch on & tuning mode)
[38]		
[39]		
[40]	⟷	12C data bus to tuner
[41]	⟶	12C clock bus to tuner
[42]		5v Supply

TMP74C434N - 3555

(26) ←———— On screen display horizontal sync pulse
(27) ←———— On screen display vertical sync pulse
(28) ←————→ On screen display oscillator
(29) ←————→ On screen display oscillator
(30) Chassis
(31) ←————→ System oscillator 4 MHz
(32) ←————→ System oscillator 4 MHz
(33) ←———— Power on reset = low
(34) Chassis
(35) ←———— Remote control data
(36) ————→ TV transmission system PAL/SECAM select
(37) ————→ TV transmission system PAL/SECAM select
(38) ————→ TV transmission system PAL/SECAM select
(39) ————→ Serial clock bus
(40) ←————→ Serial data bus
(41) ————→ TV channel auto frequency control
(42) ←———— 5 V supply

Notes:

TMP74C434N - 3555

Control microprocessor IC

Pins
(1) ————→ TV channel tuning control
(2) ————→ Sound volume control
(3) ————→ Picture brightness control
(4) ————→ Picture colour control
(5) ————→ Picture contrast control
(6) ←————→ Keypad function scanning
(7) ←————→ Keypad function scanning
(8) ←————→ Keypad function scanning
(9) ←———— Channel tuning auto frequency control
(10) ←————→ Keypad function scanning
(11) ←————→ Keypad function scanning
(12) ←————→ Keypad function scanning
(13) ←————→ Keypad function scanning
(14) ←———— SCART socket status pin 8

(Continued opposite)

TMP74C434N - 3555

(15) ← Vertical sync pulse
(16) ← Received signal ident sync pulse
(17) → External multi tuning band select IC
(18) → External multi tuning band select IC
(19) → TV channel auto frequency control on/off
(20) → LED indicator
(21) Chassis
(22)
(23) → On screen display picture blanking
(24)
(25)
(26) ← On screen display horizontal sync pulse
(27) ← On screen display vertical sync pulse
(28) ↔ On screen display oscillator
(29) ↔ On screen display oscillator
(30) Chassis
(31) ↔ System oscillator 4 MHz
(32) ↔ System oscillator 4 MHz
(33) → Power on reset = low
(34) Chassis
(35) ← Remote control data
(36) ← SCART socket RGB blanking status pin 16
(37) → External RGB blanking on /off control
(38) → TV/external composite signal select switch
(39) → 12C serial clock bus
(40) ↔ 12C serial data bus
(41) → Power on/standby control
(42) ← 5 V supply

Notes:

TMP74C434NR - 305

TMP74C434NR - 305

Control microprocessor IC

Pins

(1) → TV channel tuning control
(2) → Sound volume control
(3) → Picture brightness control
(4) → Picture colour control
(5) → Picture contrast control
(6) ↔ Keypad function scanning
(7) ↔ Keypad function scanning
(8) ↔ Keypad function scanning
(9)
(10) ↔ Keypad function scanning
(11) ↔ Keypad function scanning
(12) ↔ Keypad function scanning
(13) ↔ Keypad function scanning
(14) ← SCART socket status pin 8
(15) ← Vertical sync pulse
(16) ← Received signal ident sync pulse
(17)
(18)
(19) ← TV channel auto frequency control
(20) → LED indicator
(21) Chassis
(22) → Satellite polarizer pulse
(23) → Red on screen display
(24) → Green on screen display
(25) → Blue on screen display
(26) ← On screen display horizontal sync pulse
(27) ← On screen display vertical sync pulse
(28) ↔ On screen display oscillator
(29) ↔ On screen display oscillator
(30) Chassis
(31) ↔ System oscillator 4 MHz
(32) ↔ System oscillator 4 MHz
(33) ← Power on reset = low
(34) Chassis
(35) ← Remote control data
(36) → Satellite auto frequency control
(37) ← SCART socket RGB blanking status pin 16

(Continued opposite)

(38) ———→ TV/AV signal source select
(39) ———→ Serial clock bus
(40) ←———→ Serial data bus
(41) ———→ Power on/standby control
(42) ←——— 5 V supply

Notes:

TMP47C634N

Control microprocessor IC

Pins
(1) ———→ Picture hue control (NTSC transmission)
(2) ———→ Picture contrast control
(3) ———→ Picture colour control
(4) ———→ Picture brightness control
(5) ———→ Sound volume control
(6) ←——— SCART socket status pin 8
(7) ———→ Signal source switching ICs
(8) ———→ LED indicator
(9) ———→ LED indicator
(10) ———→ LED indicator
(11) ←———→ Keypad function scanning
(12) ←———→ Keypad function scanning
(13) ←———→ Keypad function scanning
(14) ←———→ Keypad function scanning
(15) ←———→ Keypad function scanning
(16) ←———→ Keypad function scanning
(17) ←——— Transmission system detect
(18) ←——— Transmission system detect
(19) ———→ Transmission system enable
(20) ———→ Power on/standby control
(21) Chassis
(22) ———→ Red on screen display
(23) ———→ Green on screen display
(24) ———→ Blue on screen display
(25) ———→ On screen display picture blanking

(Continued overleaf)

TMP47C634N - 2415

(26) ←—— On screen display horizontal sync pulse
(27) ←—— On screen display vertical sync pulse
(28) ←——→ On screen display oscillator
(29) ←——→ On screen display oscillator
(30)
(31) ←——→ System oscillator 4 MHz
(32) ←——→ System oscillator 4 MHz
(33) ←—— Power on reset = low
(34) ←—— Supply failure detect, settings memorized
(35) ←—— Remote control data
(36) ←—— Received signal ident sync pulse
(37)
(38) ←—— Vertical sync pulse
(39) ——→ Teletext non interlaced scanning, 25 Hz signal
(40) ←——→ Serial data bus, tuner & teletext circuit
(41) ——→ Serial clock bus, tuner & teletext circuit
(42)

Notes:

TMP47C634N - 2415

TMP47C634N - 2416

Control microprocessor IC

Pins
(1) ——→ Super VHS source select
(2) ——→ Sound volume control
(3) ——→ Picture brightness control
(4) ——→ Picture colour control
(5) ——→ Picture contrast control
(6) ←——→ Keypad function scanning
(7) ←——→ Keypad function scanning
(8) ←——→ Keypad function scanning
(9) ←—— TV channel auto frequency control
(10) ←——→ Keypad function scanning
(11) ←——→ Keypad function scanning
(12) ←——→ Keypad function scanning

1
42
21

(Continued opposite)

TMP47C634N - 2416

(13) ←——→ Keypad function scanning
(14) ←—— SCART socket status pin 8
(15) ←—— On screen display vertical sync pulse
(16) ←—— Received signal ident sync pulse
(17) ——→ TV channel auto fine tuning control
(18) ——→ Positive video modulation = high, negative = low
(19) N/c
(20) ——→ Standby LED indicator
(21) Chassis
(22) ——→ Red on screen display
(23) ——→ Green on screen display
(24) ——→ Blue on screen display
(25) ——→ On screen picture blanking
(26) ←—— On screen display horizontal sync pulse
(27) ←—— On screen display vertical sync pulse
(28) ←——→ On screen display oscillator
(29) ←——→ On screen display oscillator
(30) Chassis
(31) ←——→ System oscillator 4 MHz
(32) ←——→ System oscillator 4 MHz
(33) ←—— Power on reset = low
(34)
(35) ←—— Remote control data
(36) ←—— SCART socket RGB blanking status pin 16
(37) ——→ External RGB blanking on/off select
(38) ——→ External signal source select switching
(39) ——→ Serial clock bus
(40) ←——→ Serial data bus
(41) ——→ Power on/standby control
(42) ←—— 5 V supply

Notes:

TMP47C634NR - 331

Control microprocessor IC

Pins
(1) ⟶ Picture sharpness control
(2) ⟶ Picture contrast control
(3) ⟶ Picture colour control
(4) ⟶ Picture brightness control
(5) ⟶ Sound volume control
(6)
(7)
(8) ⟷ Keypad function scanning
(9)
10)
(11) ⟷ Keypad function scanning
(12) ⟷ Keypad function scanning
(13) ⟷ Keypad function scanning
(14) ⟷ Keypad function scanning
(15) ⟷ Keypad function scanning
(16) ⟷ Keypad function scanning
(17)
(18)
(19)
(20) ⟶ Power on/standby control = low
(21) Chassis
(22) ⟶ Red on screen display
(23) ⟶ Green on screen display
(24) ⟶ Blue on screen display
(25) ⟶ On screen display picture blanking
(26) ⟵ On screen display horizontal sync pulse
(27) ⟵ On screen display vertical sync pulse
(28) ⟷ On screen display oscillator
(29) ⟷ On screen display oscillator
(30)
(31) ⟷ System oscillator 4 MHz
(32) ⟷ System oscillator 4 MHz
(33) ⟵ Power on reset = low
(34) ⟵ Supply failure detect, settings memorized
(35) ⟵ Remote control data
(36) ⟵ Received signal ident sync pulse

(Continued opposite)

(37) ———→ Sync separator circuit mute (during switch on & tuning mode)
(38)
(39)
(40) ←——→ 12C data bus to tuner
(41) ———→ 12C clock bus to tuner
(42) ←——— 5 V supply

Notes:

TMP47C634NR - 437

Control microprocessor IC

Pins
(1) ———→ TV channel tuning control
(2) ———→ Picture contrast control
(3) ———→ Picture brightness control
(4) ———→ Picture colour control
(5) ———→ Sound volume control
(6) ———→ TV tuning band VHF H select
(7) ———→ TV tuning band VHF L select
(8) ———→ TV tuning band UHF select
(9) ←——— TV channel auto fine tuning control
(10) ←——→ System clock, external memory IC & keypad function scanning
(11) ←——→ System data, external memory IC & keypad function scanning
(12) ←——→ Keypad function scanning & external memory IC
(13) ←——→ Keypad function scanning
(14) ←——→ Keypad function scanning
(15) ←——→ Keypad function scanning
(16) ←——→ Keypad function scanning
(17) ←——→ Keypad function scanning
(18) ←——→ Keypad function scanning
(19) ———→ External memory IC select, data transfer = low
(20) ———→ Timer function LED indicator
(21) Chassis
(22) ———→ Power on/standby control

(Continued overleaf)

TMP47C634NR - 437

(23) ——→ Green on screen display
(24) ——→ Red on screen display
(25) ——→ Blue on screen display
(26) ←—— On screen display horizontal sync pulse
(27) ←—— On screen display vertical sync pulse
(28) ←——→ On screen display oscillator
(29) ←——→ On screen display oscillator
(30) Chassis
(31) ←——→ System oscillator 4 MHz
(32) ←——→ System oscillator 4 MHz
(33) ←—— Power on reset = low
(34) ←—— SCART socket status pin 8
(35) ←—— Remote control data
(36) ←—— Received signal ident sync pulse
(37) ——→ AV1/TV signal source selection
(38) ——→ AV2/TV signal source selection
(39) ——→ NICAM mono/stereo modes function select
(40) ——→ NICAM mono/stereo modes function select
(41) Chassis
(42) ←—— 5 V supply

Variation

(39), (40) Chassis

Notes:

TMP47C1237

Control microprocessor IC

Pins

(1) ⟶ TV channel tuning control
(2) ⟶ Sound balance control
(3)
(4) ⟶ TV tuning band UHF select
(5)
(6) ⟶ Positive/negative picture modulation select
(7) ⟶ SECAM L/Multi-standard transmission system select
(8) ⟶ PAL/SECAM transmission system select
(9)
(10) ⟶ TV tuning band VHF H select
(11) ⟶ TV tuning band VHF L select
(12) ⟶ Picture in picture mode (2) switching
(13) ⟵ TV channel auto frequency control
(14) ⟷ Keypad function scanning
(15) ⟵ SCART socket 1 status pin 8
(16) ⟵ Received signal ident sync pulse
(17) ⟶ Power on/standby control
(18) ⟶ LED indicator
(19) ⟶ Picture in picture mode 1 switching
(20) ⟶ External RGB blanking on/off switch
(21)
(22) ⟶ Red on screen display
(23) ⟶ Green on screen display
(24) ⟶ Blue on screen display
(25) ⟶ On screen display picture blanking
(26) ⟵ On screen display horizontal sync pulse
(27) ⟵ On screen display vertical sync pulse
(28) ⟷ On screen display oscillator
(29) ⟷ On screen display oscillator
(30) Chassis
(31) ⟷ System oscillator 4 MHz
(32) ⟷ System oscillator 4 MHz
(33) ⟵ Power on reset = low
(34) ⟵ SCART socket 2 status pin 8
(35) ⟵ Remote control data

(Continued overleaf)

TMP47C1637

(36) ⟶ AV signal source selection switching
(37) ⟶ AV signal source selection switching
(38) ⟶ AV signal source selection switching
(39) ⟶ Serial clock bus
(40) ⟷ Serial data bus
(41) ⟶ TV channel auto frequency control
(42) ⟵ 5 V supply

Notes:

TMP47C1637

Control microprocessor IC

Pins
(1) ⟶ TV channel tuning control
(2) ⟶ Picture contrast control
(3) ⟶ Picture brightness control
(4) ⟵ Headphone inserted detector
(5) ⟶ Picture colour control
(6) ⟶ Picture sharpness control
(7) ⟶ Bleep tone to sound output circuit
(8) ⟶ Signal noise reduction control
(9) ⟷ Serial data
(10) ⟵ SCART socket 1 status pin 8
(11) ⟵ SCART socket 2 status pin 8
(12) ⟶ TV/AV select
(13) ⟵ TV channel auto frequency control
(14) ⟷ Keypad function scanning
(15) ⟷ Keypad function scanning
(16) ⟷ Keypad function scanning
(17) ⟶ Power on/standby control
(18) ⟶ LED indicator
(19) ⟷ Keypad function scanning
(20)
(21) Chassis
(22) ⟶ Blue on screen display
(23) ⟶ Green on screen display
(24) ⟶ Red on screen display

(Continued opposite)

(25)⟶ On screen display picture blanking
(26)⟵ On screen display horizontal sync pulse
(27)⟵ On screen display vertical sync pulse
(28)⟷ On screen display oscillator
(29)⟷ On screen display oscillator
(30) Chassis
(31)⟷ System oscillator
(32)⟷ System oscillator
(33)⟵ Power on reset = low
(34)⟵ Received signal ident sync pulse
(35)⟵ Remote control data
(36)⟶ 50 Hz/60 Hz system scan switching
(37)⟷ Data external memory IC
(38)⟷ Data external memory IC
(39)⟷ Data external memory IC
(40)⟶ Clock bus
(41)⟶ External memory IC select, data transfer = low
(42)⟵ 5 V supply

Notes:

TMS3757 - ANL

Control microprocessor IC

Pins
(1)
(2)
(3) ⟶ TV channel tuning control
(4)
(5) Chassis
(6) ⟶ Sound volume control
(7) ⟶ Picture brightness control
(8) ⟶ Picture colour control
(9) ⟶ Picture contrast control

(Continued overleaf)

TMS3457N2L

(10) → TV/AV select
(11) N/c
(12) N/c
(13) → 12C serial clock bus
(14) ↔ 12C serial data bus
(15) N/c
(16) → Sound & picture mute
(17) N/c
(18) ← SCART socket status pin 8
(19) N/c
(20) → TV tuning band UHF/ VHF L/ VHF H select
(21) → TV tuning band UHF/ VHF L/ VHF H select
(22) → TV tuning band UHF/ VHF L/ VHF H select
(23) ← Pre-scaler, tuner oscillator sample 64
(24) ← 5 V supply
(25)
(26) ↔ System oscillator 4 MHz
(27) ↔ System oscillator 4Mhz
(28)

Notes:

TMS3457N2L

Control microprocessor IC

Pins
(1) ← –ve 32 V (program preset mode)
(2) → Auto frequency control on/off switch
(3) → TV channel auto fine tuning
(4) ↔ Keypad function scanning
(5) ↔ Keypad function scanning
(6) ↔ Keypad function scanning
(7) → TV tuning band VHF H

(Continued opposite)

(8) ————→ TV tuning band VHF L
(9) ————→ TV channel tuning control down
(10)————→ TV channel tuning control up
(11)←———— Power on reset = low
(12)←———→ System oscillator
(13)←———→ System oscillator
(14) Chassis
(15)←———— Auto frequency control
(16) N/c
(17) N/c
(18)←———→ Keypad function scanning & seven segment LED display
(19)←———→ Seven segment LED display
(20)←———→ Keypad function scanning & seven segment LED display
(21)←———→ Seven segment LED display
(22)←———→ Keypad function scanning & seven segment LED display
(23)←———→ Seven segment LED display
(24)←———→ Keypad function scanning & seven segment LED display
(25)←———→ Seven segment LED display
(26) N/c
(27)←———→ Seven segment LED display
(28)←———— 10 V supply

Notes:

TPU2734

Control microprocessor IC

Pins
(1) ←——→ System oscillator 4 MHz
(2) Chassis
(3) N/c
(4) ←—— Power on reset = low
(5) ←—— Power switch pulse contact
(6) Chassis
(7) ←——→ IM bus data
(8) ——→ IM bus ident/control
(9) ——→ IM bus clock
(10)——→ TV channel tuning up
(11)——→ TV channel tuning down
(12)←—— Remote control data
(13)←—— Tuner oscillator sample 64, pre-scaler
(14)←——→ Keypad function scanning & seven segment LED display
(15)←——→ Keypad function scanning & seven segment LED display
(16)←——→ Keypad function scanning & seven segment LED display
(17)←——→ Keypad function scanning & seven segment LED display
(18)←——→ Keypad function scanning & seven segment LED display
(19)←——→ Keypad function scanning & seven segment LED display
(20) Chassis
(21)←——→ Keypad function scanning & seven segment LED display
(22)←——→ Keypad function scanning & seven segment LED display
(23)←——→ SCART socket status pin 8 & remote data
(24)←—— External ICs reset
(25)——→ Picture in picture ICs enable
(26)←——→ Keypad function scanning & seven segment LED display
(27)←—— 5 V supply
(28)←——→ 12C bus data

(Continued opposite)

(29) → 12C bus clock
(30) → TV tuning band VHF/UHF select
(31) → TV tuning band VHF/UHF select
(32) → TV/AV select
(33)
(34) ← SCART socket status pin 8
(35) → LED indicator driver
(36) → Seven segment LED display supply
(37) → Seven segment LED display supply
(38)
(39) → Sound bleep to sound output circuit
(40) ← 5 V supply

Notes:

TVP02066

Control microprocessor IC

Pins
(1) ← On screen display vertical sync pulse
(2) ← On screen display horizontal sync pulse
(3) → Picture colour control
(4) → Sound volume control
(5) ← TV channel auto frequency control
(6) ↔ Keypad function scanning
(7) ↔ Keypad function scanning
(8) ← Horizontal sync pulse
(9) ← 5 V supply
(10) Chassis
(11) ← 5 V supply
(12) ↔ System oscillator 4 MHz
(13) ← Received signal ident sync pulse
(14) ↔ External memory IC
(15) ← Power on reset = low
(16) → Power on/standby control
(17) Chassis
(18) ↔ IM bus data
(19) → IM bus ident/control

(Continued overleaf)

TVP02066

(20) ⟶ IM bus clock
(21) ⟶ Picture brightness control
(22) ⟶ Picture contrast control
(23) ⟵ Remote control data
(24) ⟶ TV channel tuning control
(25) ⟷ Keypad function scanning
(26) ⟷ Keypad function scanning
(27) ⟷ Keypad function scanning
(28) ⟷ Keypad function scanning
(29) ⟶ TV tuning band VHF/UHF select
(30) ⟶ TV tuning band VHF/UHF select
(31) ⟶ AV signal source select
(32) ⟶ Transmission system PAL/SECAM select
(33) ⟶ TV/external RGB/Super VHS select
(34) ⟶ TV/AV signal source select
(35) ⟶ Picture mute
(36) ⟷ Keypad function scanning
(37) ⟶ Red on screen display
(38) ⟶ Green on screen display
(39) ⟶ Blue on screen display
(40) ⟶ On screen display picture blanking

Notes:

UPD1514C

μPD1514C

Control microprocessor IC

Pins
(1) ——→ Power / standby control
(2) ——→ Mute LED indicator
(3) ——→ Serial clock bus
(4) ←——→ Data, teletext control & keypad function scanning
(5) ←——→ Preset sound volume & keypad function scanning
(6) ——→ Command received LED indicator
(7) ←——→ Keypad function scanning
(8) ←——→ Keypad function scanning
(9) ←——→ Keypad function scanning
(10) ←——→ Keypad function scanning
(11) ←—— Remote control data
(12) ←—— Power switch pulse contact
(13) Chassis
(14) Chassis
(15) ←——→ System oscillator 3.58 MHz
(16) ←——→ System oscillator 3.58 MHz
(17) ←—— Power on reset = low
(18) ——→ Picture colour increase control
(19) ——→ Picture colour decrease control
(20) ——→ Picture brightness increase control
(21) ——→ Picture brightness decrease control
(22) ——→ Sound volume increase control
(23) ——→ Sound volume decrease control
(24) ——→ Initial switch on volume control
(25) ——→ TV / teletext select
(26) ——→ TV channel tuning control up
(27) ——→ TV channel tuning control down
(28) ←—— 5 V supply

Notes:

UPD6124G

UPD6124G

μPD6124G

Infra-red remote controller IC

Pins
(1) ←——→ Keypad functions scanning
(2) ←——→ Keypad functions scanning
(3) N/c
(4) N/c
(5) ——→ Infra-red transmitter diode driver
(6) ←—— 3 V supply
(7) ←——→ System oscillator 455 kHz
(8) ←——→ System oscillator 455 kHz
(9) Ground
(10)
(11) ←——→ Keypad functions scanning
(12) ←——→ Keypad functions scanning
(13) ←——→ Keypad functions scanning
(14) ←——→ Keypad functions scanning
(15) ←——→ Keypad functions scanning
(16) ←——→ Keypad functions scanning
(17) ←——→ Keypad functions scanning
(18) ←——→ Keypad functions scanning
(19) ←——→ Keypad functions scanning
(20) ←——→ Keypad functions scanning

Notes:

UPD6125G-504 or μPD6125G-504

UPD6125G-504 or μPD6125G-504

UPD6125AG-561 or μPD6125AG-561

UPD6125AG-578 or μPD6125AG-578

Infra-red remote controller IC

Pins
(1) ←——→ Keypad functions scanning
(2) ←——→ Keypad functions scanning
(3) ←——→ Keypad functions scanning
(4) ←——→ Keypad functions scanning
(5) N/c
(6) N/c
(7) ——→ Infra-red transmitting diode driver
(8) ←—— 3 V supply
(9) ←——→ System oscillator
(10) ←——→ System oscillator
(11) Ground
(12)
(13) ←——→ Keypad functions scanning
(14) ←——→ Keypad functions scanning
(15) ←——→ Keypad functions scanning
(16) ←——→ Keypad functions scanning
(17) ←——→ Keypad functions scanning
(18) ←——→ Keypad functions scanning
(19) ←——→ Keypad functions scanning
(20) ←——→ Keypad functions scanning
(21) ←——→ Keypad functions scanning
(22) ←——→ Keypad functions scanning
(23) ←——→ Keypad functions scanning
(24) ←——→ Keypad functions scanning

Variations

Pins
(5) ←——→ Keypad functions scanning

Notes:

UPD6125CA604

UPD6125CA604

µPD6125CA604

Infra-red remote controller IC

Pin
(1) ←——→ Keypad functions scanning
(2) ←——→ Keypad functions scanning
(3) ←——→ Keypad functions scanning
(4) ←——→ Keypad functions scanning
(5) ←——→ Keypad functions scanning
(6) ——→ Infra - red transmitting diode driver
(7) N/c
(8) ←—— 6 V supply
(9) ←——→ System oscillator 455 kHz
(10) ←——→ System oscillator 455 kHz
(11) Ground
(12)
(13) ←——→ Keypad function scanning
(14) ←——→ Keypad function scanning
(15) ←——→ Keypad function scanning
(16) ←——→ Keypad function scanning
(17) ←——→ Keypad function scanning
(18) ←——→ Keypad function scanning
(19) ←——→ Keypad function scanning
(20) ←——→ Keypad function scanning
(21) ←——→ Keypad function scanning
(22) ←——→ Keypad function scanning
(23) N/c
(24) ←——→ LED display driver

1 24 12

Notes:

UPD6600AGS-B78

μPD6600AGS-B78

Infra-red remote controller IC

Pins

(1) ←→ Keypad functions scanning
(2) ←→ Keypad functions scanning
(3) ←→ Keypad functions scanning
(4) N/c
(5) → Infra-red transmitting diode driver
(6) ← 3 V supply
(7) ←→ System oscillator 455kHz
(8) ←→ System oscillator 455kHz
(9) Ground
(10)
(11) ←→ Keypad functions scanning
(12) ←→ Keypad functions scanning
(13) ←→ Keypad functions scanning
(14) ←→ Keypad functions scanning
(15) ←→ Keypad functions scanning
(16) ←→ Keypad functions scanning
(17) ←→ Keypad functions scanning
(18) ←→ Keypad functions scanning
(19) ←→ Keypad functions scanning
(20) ←→ Keypad functions scanning

Notes:

UPD6600GS-57

UPD6600GS-57

μPD6600GS-57

Infra-red remote controller IC

Pins
(1) ←——→ Keypad functions scanning
(2) ←——→ Keypad functions scanning
(3) N/c
(4) N/c
(5) ——→ Infra-red transmitting diode driver
(6) ←—— 3 V supply
(7) ←——→ System oscillator 455kHz
(8) ←——→ System oscillator 455kHz
(9) Ground
(10)
(11) ←——→ Keypad functions scanning
(12) ←——→ Keypad functions scanning
(13) ←——→ Keypad functions scanning
(14) ←——→ Keypad functions scanning
(15) ←——→ Keypad functions scanning
(16) ←——→ Keypad functions scanning
(17) ←——→ Keypad functions scanning
(18) ←——→ Keypad functions scanning
(19) ←——→ Keypad functions scanning
(20) ←——→ Keypad functions scanning

1 20 10

Notes:

XC86627P

Control microprocessor IC

Pins
(1) Chassis
(2) ←——— Power on reset = low
(3) ←——— Remote control data
(4) ←——— 5 V supply
(5) ←———→ System oscillator 4Mhz
(6) ←———→ System oscillator 4Mhz
(7) ←——— 5 V supply
(8) N/c
(9) ———→ SECAM reception LED indicator
(10) ———→ NTSC reception LED indicator
(11) ———→ PAL reception LED indicator
(12)
(13) ———→ Stereo reception LED indicator
(14) ←——— Received signal ident sync pulse
(15) ←——— Remote control serial clock bus
(16) ←———→ NICAM circuit serial clock bus
(17) ←———→ Teletext circuit serial data bus
(18) ———→ Teletext circuit serial clock bus
(19) ———→ PAL transmission colour system = high
(20) ———→ SECAM transmission system = high
(21) ←———→ Serial data bus, external memory IC & circuits
(22) ———→ TV I.F enable
(23) ———→ Serial clock bus external memory & circuits
(24) ———→ Serial clock bus, teletext & AV circuits ICs
(25) ←———→ Keypad functions scanning & seven segment display
(26) ←———→ Keypad functions scanning & seven segment display
(27) ←———→ Keypad functions scanning & seven segment display
(28) ←———→ Keypad functions scanning & seven segment display
(29) ←———→ Keypad functions scanning & seven segment display
(30) ←———→ Keypad functions scanning & seven segment display
(31) ←———→ Keypad functions scanning & seven segment display
(32) ←———→ Keypad functions scanning & seven segment display
(33) ←———→ Keypad functions scanning & seven segment display
(34) ←———→ Keypad functions scanning & seven segment display
(35) ←———→ Keypad functions scanning & seven segment display
(36) ←———→ Keypad functions scanning & seven segment display
(37) ←———→ Keypad functions scanning & seven segment display

(Continued overleaf)

(38) ⟷ Keypad functions scanning & seven segment display
(39) ⟵ SCART socket status pin 8
(40) ⟶ Teletext enable

Notes:

ZC88604

ZC88606

ZC88619

ZC88621

Control microprocessor IC

Pins
(1) ⟵ Power on reset = low
(2) N/c
(3) ⟷ Data bus
(4) ⟶ Clock bus
(5) N/c
(6) ⟵ SCART socket status pin 8
(7) N/c
(8) ⟵ Protection circuit sensing
(9) ⟶ TV/AV select
(10)
(12) N/c
(13) ⟵ Received signal ident sync pulse
(14) ⟶ Power on/standby control
(15) ⟶ Power switch pulse contact
(16) ⟶ Blue on screen display
(17) ⟶ Green on screen display
(18) ⟶ Red on screen display
(19) ⟵ 5 V supply
(20) Chassis
(21) ⟶ On screen display picture blanking
(22)
(23) ⟵ On screen display horizontal sync pulse
(24) ⟵ On screen display vertical sync pulse

(Continued opposite)

(25) N/c
(26)
(27) ←——→ Keypad function scanning
(28) ←——→ Keypad function scanning
(29) ←——→ Keypad function scanning
(30) ←——→ Keypad function scanning
(31) ——→ Picture contrast control
(32) ——→ Picture colour control
(33) ——→ Picture brightness control
(34) ——→ Sound balance control
(35) ——→ Sound volume control
(36) Chassis
(37) N/c
(38) ←—— Remote control data
(39) ←——→ System oscillator 4Mhz
(40) ←——→ System oscillator 4 MHz

Notes:

ZC411856P

Control microprocessor IC

Pins
(1) ——→ Power on reset = low
(2)
(3) ←—— Vertical sync pulse
(4) ——→ Clock bus
(5) ——→ TV select = high
(6) ←—— SCART socket status pin 8
(7) ——→ TV I.F enable
(8) ←—— Beam current protection circuit
(9) ——→ TV/AV select
(10) Chassis
(11) Chassis
(12) Chassis
(13) ←—— Received signal ident sync pulse
(14) ——→ Power on/standby control
(15) ——→ Power switch pulse contact

(Continued overleaf)

1 40 20

ZC411856P

(16) → Blue on screen display
(17) → Green on screen display
(18) → Red on screen display
(19) ← 5 V supply
(20) Chassis
(21) → On screen display picture blanking
(22) → Tube degauss circuit on = high
(23) ← On screen display horizontal sync pulse
(24) ← On screen display vertical sync pulse
(25)
(26)
(27) ↔ Keypad function scanning
(28) ↔ Keypad function scanning
(29) ↔ Keypad function scanning
(30) ↔ Keypad function scanning
(31) → Picture contrast control
(32) → Picture colour control
(33) → Picture brightness control
(34) → Sound balance control
(35) → Sound volume control
(36) Chassis
(37) ← Remote control data
(38) ↔ Data
(39) ↔ System oscillator 4 MHz
(40) ↔ System oscillator 4 MHz

Notes:

6805T2

Control microprocessor IC

Pins
(1) Chassis
(2) ←——— Remote control data
(3) ←——— 5 V supply
(4) ←———→ System oscillator 4MHz
(5) ←———→ System oscillator 4 MHz
(6)
(7) ———→ TV channel tuning control
(8)
(9)
(10) ———→ Power on/standby control
(11) ←——— Tuner oscillator sample 64, pre-scaler
(12) ———→ Seven segment LED display
(13) ———→ Seven segment LED display
(14) ———→ Seven segment LED display
(15) ———→ Seven segment LED display
(16) ———→ Seven segment LED display
(17) ———→ Seven segment LED display
(18) ———→ Seven segment LED display
(19) ———→ Seven segment LED display
(20)
(21) ———→ Seven segment LED display supply
(22) ———→ Seven segment LED display supply
(23) ———→ Clock, external memory IC
(24) ———→ Clock, digital to analogue converter & tuning control ICs
(25) ←———→ Data, external memory & digital to analogue converter ICs
(26) ———→ External memory IC select, data transfer = low
(27) ←——— Received signal ident sync pulse/mute
(28) ←——— Power on reset = low

Notes: